Geological Museum Institute of Geological Scienc

The story of the Earth

The Earth is a small, rocky planet with large quantities of surface water whose average temperature lies between freezing point and boiling point.
It also has a dense oxygen-rich atmosphere and supports life in abundance.

Yet the cool blue-and-white face it presents to outer space conceals a fiery white-hot interior. The cosmic origin of this ceaselessly moving inferno and how it has created the environment of land, sea and air we live in is told here.

CONTENTS

London: Her Majesty's Stationery Office

Planet Earth

The Earth in Space

The Earth, our planet, seems large to us. Its radius of 6400 km and circumference of 40 000 km are enormous compared to distances we are used to. But when compared to the size of the Universe, the Earth and Solar System are infinitesimally small.

Astronomical distances are too great for the kilometre to be used as a unit; they are expressed in terms of the *light-year*, the distance light travels in one year at 300 000 km per second, that is almost 10 million million kilometres.

The Moon is 1.25 light-seconds from the Earth, which itself is 8 light-minutes from the Sun. The average diameter of the Solar System is 11 light-hours. These distances are small compared to the 4 light-years which separate us from α-Centauri, the nearest of the 100 thousand million stars which make up our Galaxy, but large compared to the 0.025 light-seconds which separate London from New York. If the orbit of the Earth were the size of a 5p coin in the Geological Museum in South Kensington, then α-Centauri would be situated in Piccadilly Circus.

Our Galaxy is a flattened disc of gas, dust and stars 80 000 light-years across, made up of two tightly spiralling arms. Our view of the Galaxy in the sky we call the Milky Way.

Our Galaxy is only one of millions. It is a member of a cluster of more than twenty galaxies of assorted shapes and sizes scattered over two or three million light-years of space around us. As far as telescopes can reach, Man can detect galaxies and clusters of galaxies evenly distributed in all directions; at least 400 million galaxies are detectable from Earth. An observer in the group of galaxies in Coma Berenice looking at the Earth today through an enormously powerful telescope could see the dinosaurs alive. Radiation from the most distant sources 8000 million light-years away was emitted long before the formation of the Earth and many of these sources may no longer exist.

Origin of the elements

Elements are primary, chemically indivisible substances. Of all the known elements at least 91 occur naturally on Earth. Over 70 have been spectroscopically identified in the Sun. Their origin, relative abundance and distribution in the Universe are now known to be closely associated with the evolution and life histories of stars. Astronomical observations and laboratory researches indicate that thermonuclear reactions within the intensely hot interiors of stars are able to create, step by step, all the elements from primordial hydrogen, which is the simplest, lightest and most abundant element in the Universe. In brief, the history of a star is broadly defined by successive stages of internal gravitational contraction and thermonuclear 'burning'. Contraction causes heating up of the star's core to temperature ranges through which its atomic fuel is transformed, by various thermonuclear processes involving protons (hydrogen nuclei), neutrons and nuclei of light elements, into heavier elements, releasing radiant energy (mostly as light). Many old stars explode (as *novae* and *supernovae*) scattering their element stockpiles into space. New stars, partly composed of this earlier-formed element debris, are able to extend the nuclear reaction processes (if their sizes and temperatures are adequate) to synthesize elements of progressively higher atomic weight. Thus, all materials of the Solar System originated over 5000 million years ago in the birth and death of stars.

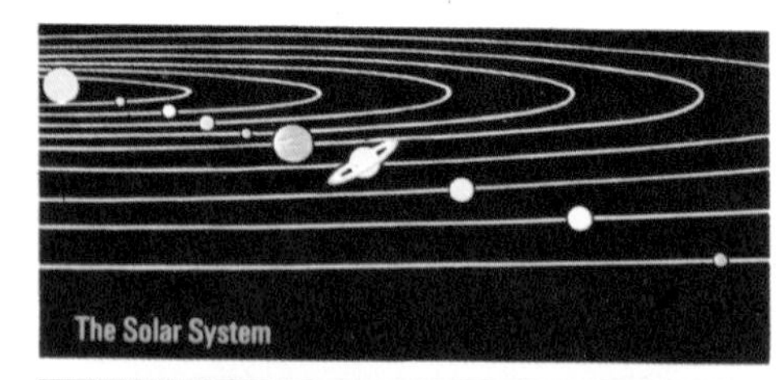

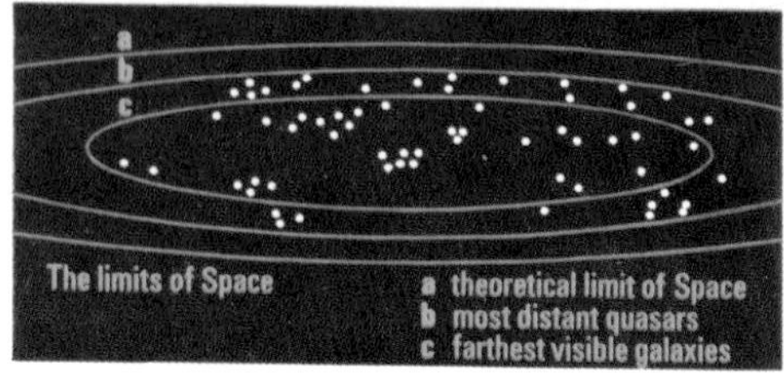

2 The Earth in Space

Origin and age of the Solar System

The Solar System consists of planets, moons, asteroids, comets, meteorites, dust and gas, and a central star – the Sun – about which all these revolve. The fact that most of this material circles the Sun in the same direction, mostly spinning the same way and within the same plane, leaves little doubt that it nearly all formed together. Different parts of the Solar System should, therefore, have similar ages: the oldest Moon rocks measured have ages of around 4600 million years; stony meteorites have been dated at 4550 million years; and the age of the Earth's crust has been calculated as 4550 million years. The close agreement of these ages suggests that the Solar System is about 4600 million years old.

Exactly how the Solar System formed is not yet known. Most ideas are based on the theory that the whole Solar System formed together from part of a vast, contracting cloud of gas and dust. New stars were created wherever the cloud was compressed to thermonuclear fusion temperatures. In this way, the Sun was born within a rotating disc of dust and gas (fig 3a). Out of this disc, planets could have formed, perhaps by gravitational collapse into large gas-ball protoplanets (3b right). Those nearer the Sun must then have lost their gas envelopes, leaving behind molten, rocky cores as small planets like the Earth. Alternatively, dust could have gathered to form belts of asteroids, which ultimately clumped to form molten planets, including the Earth (3b left). It is unlikely, however, that the Solar System had a simple origin. The outer planets may well have formed from protoplanet discs complete with 'mini-planet' systems of moons. The inner planets may have formed later, from dust rings in the Solar disc, perhaps with some direct condensation from hot Solar vapour.

3 Origin of the Solar System

The Sun and Planets

The Sun is an average-size star mostly composed of ultra-high-temperature hydrogen and helium. It is more than 109 times the diameter of the Earth, and its mass constitutes more than 99% of the Solar System. The Sun's energy stems from thermonuclear reactions in its core, where hydrogen is converted to helium at very high temperatures and pressures. The temperature of its visible surface is about 6000°C.

Mercury is small (diameter 4880 km). Its density is high (5.4 g/cc) and it probably has a large iron-rich core with overlying mantle and crust of iron-silicate rocks: the surface is cratered like the Moon's. Surface temperatures range from 500°C at noon to −180°C at night.

Venus is Earth-like in size (12104 km in diameter), and with a density of 5.2 g/cc is probably similar in composition and structure. Its atmosphere, largely carbon dioxide, is 100 times as dense as Earth's, so its surface temperature, about 480°C, varies little.

Earth is the largest of the 'terrestrial' planets (diameter 12756 km, density 5.5 g/cc). It is layered, with a dense iron-rich core surrounded by mantle and thin crust of silicate rocks. The crust is seven-tenths covered by water, which, with the atmosphere of oxygen and nitrogen, supports life. Surface temperatures range from 60°C to −90°C. It has a single large moon.

Mars is much smaller than Earth (diameter 6787 km, density 3.9 g/cc), but is also layered into core, mantle and crust. A thin atmosphere mostly consists of carbon dioxide; ice and frozen carbon dioxide form seasonally variable polar icecaps. The freezing lifeless surface is cratered like the Moon but has volcanoes and erosion features. Mars has two small moons.

The Asteroids are the many thousands of small 'minor planets', most of which orbit the Sun between Mars and Jupiter. Most asteroids are rocky and less than 10 km across; a few exceed 400 km and the largest (Ceres) is 955 km in diameter. Some asteroid collision fragments may have been perturbed by Jupiter into orbits that bring them to Earth as meteorites.

Jupiter is the largest planet in the Solar System (diameter 142800 km), with 17 known moons. It spins rapidly, completing one rotation in under 10 hours. Jupiter's density is low (1.3 g/cc); a thick atmosphere of hydrogen, helium and ammonia passes down into a planetary body largely composed of liquid and solid hydrogen. The Great Red Spot (fig 4) is a permanent high-pressure atmospheric storm.

Saturn is a huge planet (diameter 120000 km) with 15 moons. Its distinctive rings are probably formed by myriads of small ice-covered particles. Like Jupiter, Saturn consists largely of hydrogen, with some helium and methane in its atmosphere: an average piece of Saturn (density 0.7 g/cc) would float on water. Its moon Titan is enshrouded in nitrogen.

Uranus is a large planet (diameter 51800 km) with five known satellites and a faint ring system. Unlike all other planets Uranus lies on its side, rotating in the plane of its orbit around the Sun. Mostly composed of liquid and solid hydrogen, giving it a density of 1.2g/cc, Uranus has an atmosphere of hydrogen, helium and methane which appears greenish.

Neptune is smaller than Uranus (diameter 49500 km) and although more massive (density 1.7 g/cc) is probably similar in composition and structure. Like Uranus it has a slightly green methane-rich atmosphere. It has two moons.

Pluto is the most enigmatic of the planets. On the assumption that its surface is largely frozen methane, its reflectance of sunlight suggests a diameter of about 3000 km. It takes 248 years to encircle the Sun and its highly elliptical orbit periodically brings it closer to the Sun than Neptune. Pluto has a single large moon.

4 Main components of the Solar System

NASA

5 Earth-rise over the Moon seen from Apollo 12, November 1969

The Moon, planet Earth's natural satellite, circles its partner at an average distance of 384 000 km. It is little more than a quarter the diameter of the Earth, has a lower mean density and merely a sixth of the Earth's gravity. It has no water, atmosphere or life.

Forming the Moon's surface are comparatively smooth dark areas of low relief (*maria*), and brightly reflecting rugged mountainous regions (*terrae*). Circular craters, of all sizes up to 200 km across, mostly produced by impact of countless meteorites, pockmark the landscape, larger craters being more numerous over the terrae (fig 5). The terrae are remnants of the Moon's primeval crust, which solidified from a molten surface about 4500 million years ago. The maria are later impact-excavated basins in this crust, repeatedly flooded by basaltic lavas erupted from within the Moon between 4000 and 3000 million years ago. Since then the Moon has been a virtually dead world, disturbed only by the occasional meteorite fall.

NASA

6 Surface of Mars viewed from Viking Lander 1, July 1976

Spacecraft exploration of the Solar System is yielding much new information about the planets and their satellites. The surface of Mercury is heavily impact-cratered, with large lava-filled basins, like our Moon. We now know that the cloud-hidden surface of Venus is mostly rolling plains, with some continent-size upland areas, deep rift valleys, mountains 10 km high, volcanoes and impact craters. On Mars there are vast canyons 5 km deep, a huge volcano 20 km high, impact craters and ancient water-eroded channels: we have close-up views of rock-strewn, 'sandy' deserts (fig 6).

Features of Jupiter's four large, planet-like moons include the ancient densely-cratered ice crust of Callisto, the smooth 'cracked egg' icy surface of Europa, and the youthful, almost craterless but sulphurous landscape of Io, continually modified by sulphur volcanoes in spectacular eruption. Of Saturn's five larger satellites four are impact-cratered and icy: Titan is nitrogen-enshrouded, with a surface 'ocean' probably of liquid methane.

The Earth

7 The Earth's interior

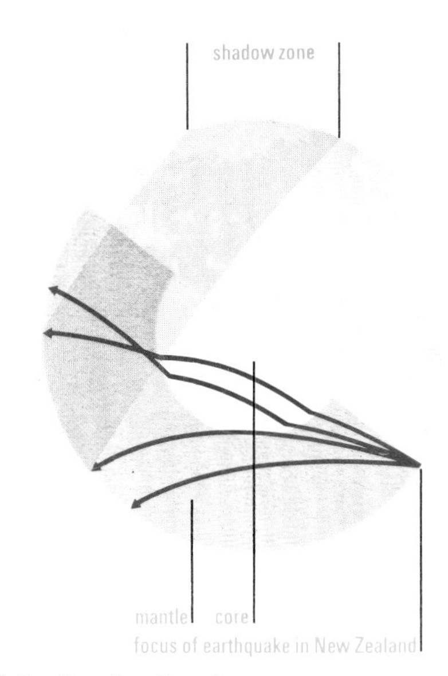

8 Bending of earthquake waves

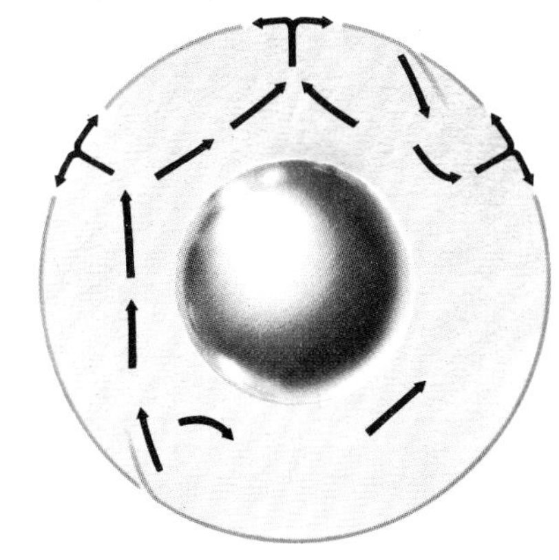
9 Irregular convection currents in the mantle

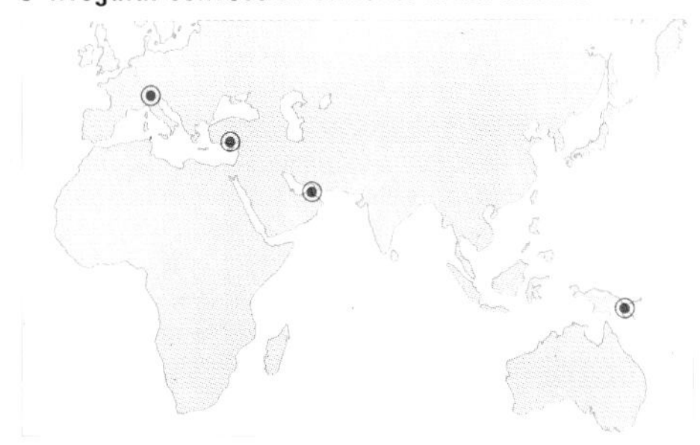
10 Outcrops of mantle material

The Earth's interior

The Earth's interior has a layered structure: the outermost layer is the *crust;* beneath is the *mantle,* and then the *core.* Scientists cannot study the interior of the Earth directly. What we know of it is worked out from studying the way earthquake waves are bent as they pass through the Earth (fig 8). There are four places on the Earth's surface, shown in fig 10, where rock thought to be mantle material crops out. This rock, which is heavy and dark in colour, is called *peridotite.* It is composed mainly of the silicate minerals olivine and pyroxene. Other likely constituents of the upper mantle are *dunite* (pure olivine rock) and *eclogite,* a dense form of basalt. At greater depths these materials would change to more dense forms and ultimately in the lower mantle probably break down to simple oxides. Movements in the mantle, presumably caused by convection, are reflected in the structure of the crust. To judge from this structure, the convection pattern must be irregular, as in fig 9. The Earth's core is a dense material, probably a mixture of iron and iron sulphide. The outer core behaves as a liquid, but the inner core is a solid. The Earth's magnetic field is generated by circulation movements in the liquid outer core.

The Earth's crust

The Earth's crust is a thin skin of rock less dense than the underlying mantle, from which it has been derived by complex processes operating over many millions of years. Earthquake waves speed up 15 per cent as they cross from the crust into the mantle. The junction between the two is called the *Mohorovičić Discontinuity,* or *M-Discontinuity* for short, after the Yugoslav scientist who discovered it.

There are two kinds of crust: *continental* and *oceanic.* Compared with the oceanic crust, the continental crust is: lighter in weight (its average density is 2.7 against 3.0 for the oceanic crust and 3.4 for the upper mantle); much thicker (it averages 35-40 km and reaches 60-70 km under high mountain chains; the oceanic crust averages only 6 km); and very much older (parts are older than 3500 million years, and very large areas are older than 1500 million years; the oceanic crust is nowhere older than 200 million years). The continental crust also has a very complicated structure and variable composition, whereas the oceanic crust has a simple layered structure and uniform composition. The junction of continental and oceanic crust is normally a fracture obscured by recent sedimentation.

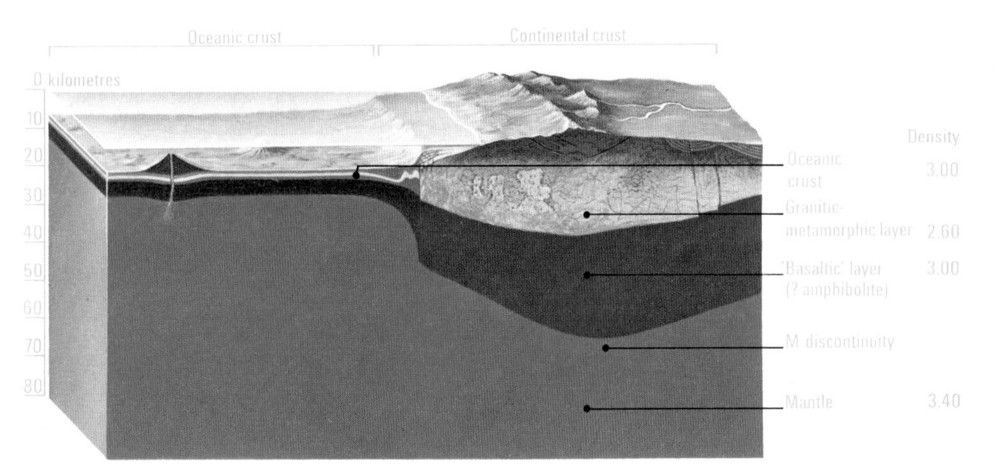

11 The two main types of crust

12 Evolution of the atmosphere

BARRY EVANS

13 Semi-molten surface of the Earth 4500 million years ago

Origin of the primitive crust, the seas and the atmosphere

The newly-formed Earth 4500 million years ago must have been very different from the Earth we know today or even that of 1000 million years ago. It may have been a much larger planet, not unlike Saturn, with a huge atmosphere of cosmic gases and a rocky core which was probably entirely molten. This primordial atmosphere was stripped away, possibly during a period of higher heat output by the Sun. We know this from the very small content of rare gases neon and xenon in our present atmosphere compared with the cosmic mixture. Our present atmosphere and surface waters were derived entirely from the Earth itself, that is, from the molten core of the primordial planet. Vast amounts of gas must have been exhaled from the semi-molten surface of the early Earth. At the same time, the primitive crust was forming. Fig 13 is an artist's reconstruction of the scene at the time. This first crust was probably of basaltic composition. As the crust repeatedly cracked and re-melted, a lighter granitic liquid would gradually have separated out. Further concentration of this granitic material into a thick granitic (continental) crust may have had to await the establishment of erosional processes involving surface water condensed from the exhaled gases. Re-melting of the products of erosion is an alternative way of producing granitic rocks. Even after the formation of a totally solid crust and the first true oceans, intense volcanic activity went on adding gases and water vapour to the atmosphere. These volcano gases were poor in oxygen, which was not added to the atmosphere until advanced forms of oxygen-producing plant life had evolved around 1900 million years ago. These were forms able to survive in the presence of oxygen produced from carbon dioxide in the process of photosynthesis. Fig 12 shows the evolution of the atmosphere in four stages: (a) primordial gases, later lost; (b) exhalations from the molten surface; (c) steady additions from volcanic activity; (d) addition of oxygen by plant life.

The oceanic crust

Our understanding of the nature and geological history of the ocean floors has greatly increased in recent years. This has been achieved by intensive exploration in which special instruments and many different techniques of investigation have been employed. In this wealth of new information the cornerstone discovery has been the comparative youth of the ocean floors (and underlying oceanic crust) relative to those parts of the Earth's crust forming the continents.

The floor of the oceans Some two-thirds of the Earth's surface lies beneath the oceans. This submarine environment ranges from lowest tidal shorelines to the very deepest parts (11 137 metres in the Marianas Trench in the Pacific) and has an average depth of 3870 metres (or $4\frac{1}{2}$ times the mean height of continents).

Seawards from the coasts, and still part of the continental crust, lie the submerged continental margins consisting of gently sloping *continental shelves* of variable width which drop more steeply down the *continental slopes* to the ocean floor at around the 200 metre contour. Much of the ocean floor lies at depths of about 5 km. Here it forms the *abyssal plains* on which submarine mountains rise as islands (some with coral reefs), drowned peaks (*seamounts*) and wave-eroded, flat-topped seamounts (*guyots*), all of which are of volcanic origin. Common to all oceans are the *ridges*, 2-4 km in height and up to 4000 km wide, which form a nearly continuous submarine mountain range over 40 000 km in length, appearing above sea level in places like Iceland and Tristan da Cunha. They are broken and repeatedly offset by innumerable *transform faults.* Deep, furrow-like *trenches* hundreds of kilometres long, tens of kilometres wide, often flanked by chains of volcanic islands (or *island arcs)*, border some oceanic regions.

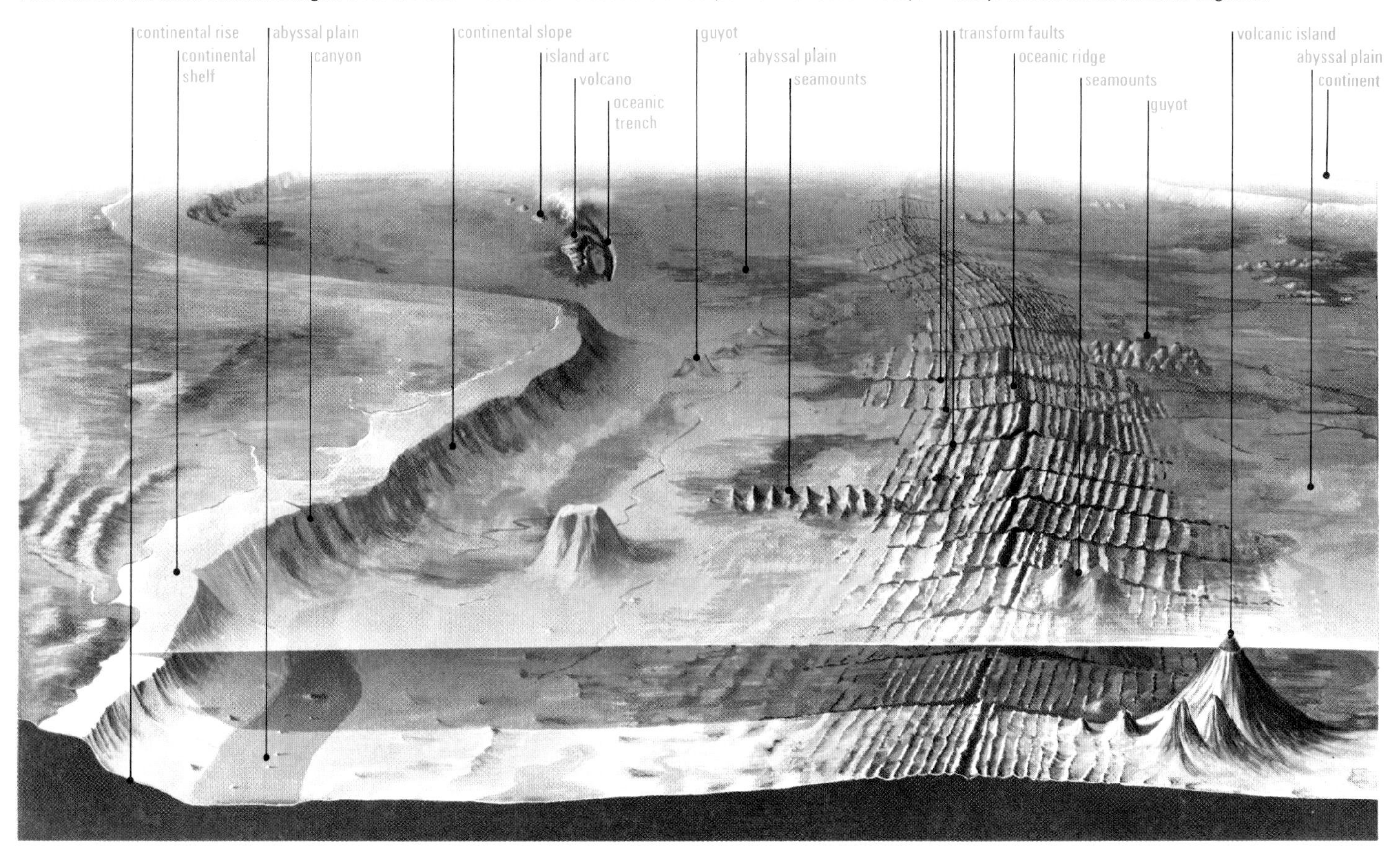

14 Main features of the ocean floor

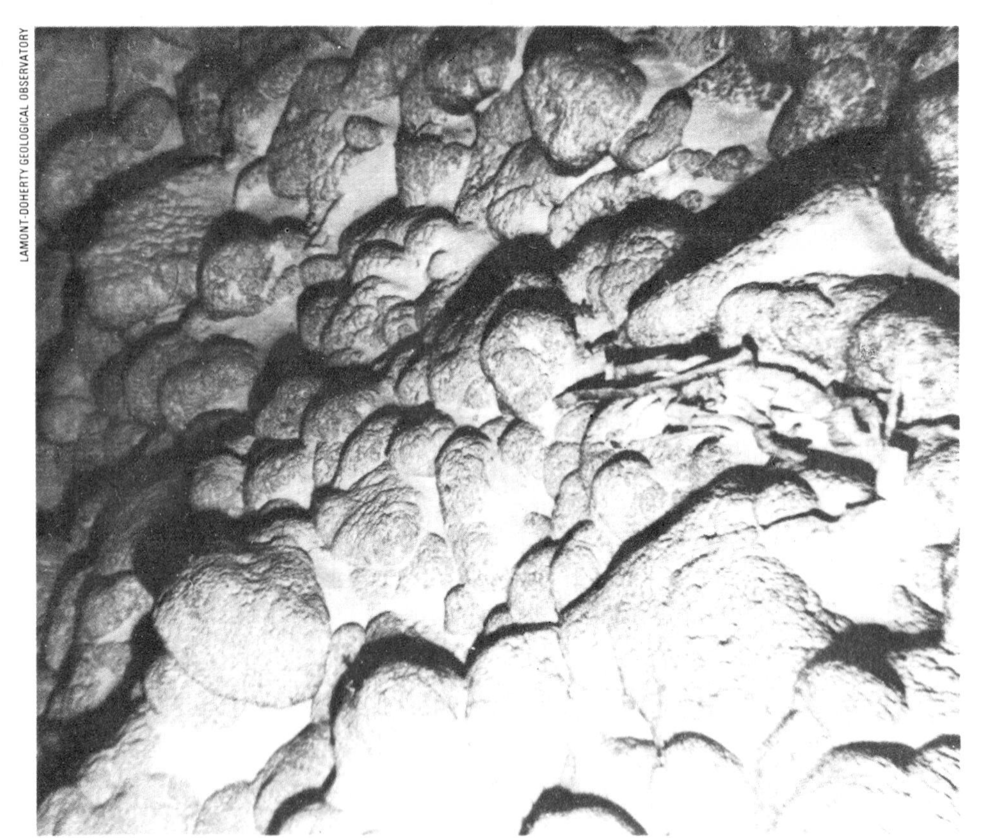

LAMONT-DOHERTY GEOLOGICAL OBSERVATORY

15 Pillow lavas on the western slope of the mid-Atlantic ridge, 2650 metres below sea level

Structure and origin of the oceanic crust

Our knowledge and ideas on the composition of the oceanic crust are based mainly on geophysical evidence, dredged rock samples, deep-sea drilling by the American research vessel *Glomar Challenger,* and comparisons with rock sequences on land thought to represent upthrust oceanic crust. The uppermost layer consists of deep-sea sediments; this layer thickens away from the mid-ocean ridges, eventually reaching more than a kilometre and spanning more than 150 million years near the ocean margins. Beneath the sediments is the heavily fractured igneous basement, composed in its upper part of basalt lava, 75 per cent of which is pillow lava erupted on the ocean floor as rounded lumps or pillows in small volcanic hills (fig 15). The lavas pass down into a layer inferred to consist of vertical sheet-like veins or dykes which are the feeders for the lava flows. This layer is considerably thicker in oceanic crust that has been formed at slow spreading rates (fig 16a). Beneath is a thick layer presumed to be gabbro, a coarsely crystalline form of basalt, the lower part of which may consist of layered cumulate gabbro formed by crystal settling. New oceanic crust is thus created by several processes – magma freezing, crystal settling, dyke injection and lava eruption –

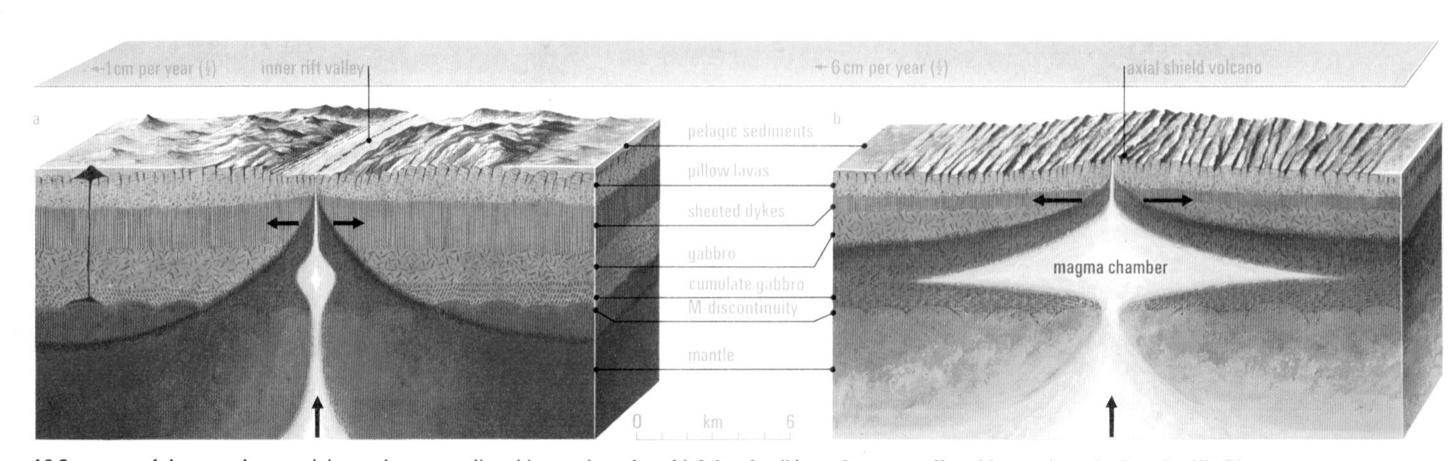

16 Structure of the oceanic crust (a) at a slow-spreading ridge such as the mid-Atlantic; (b) at a fast-spreading ridge such as the East Pacific Rise

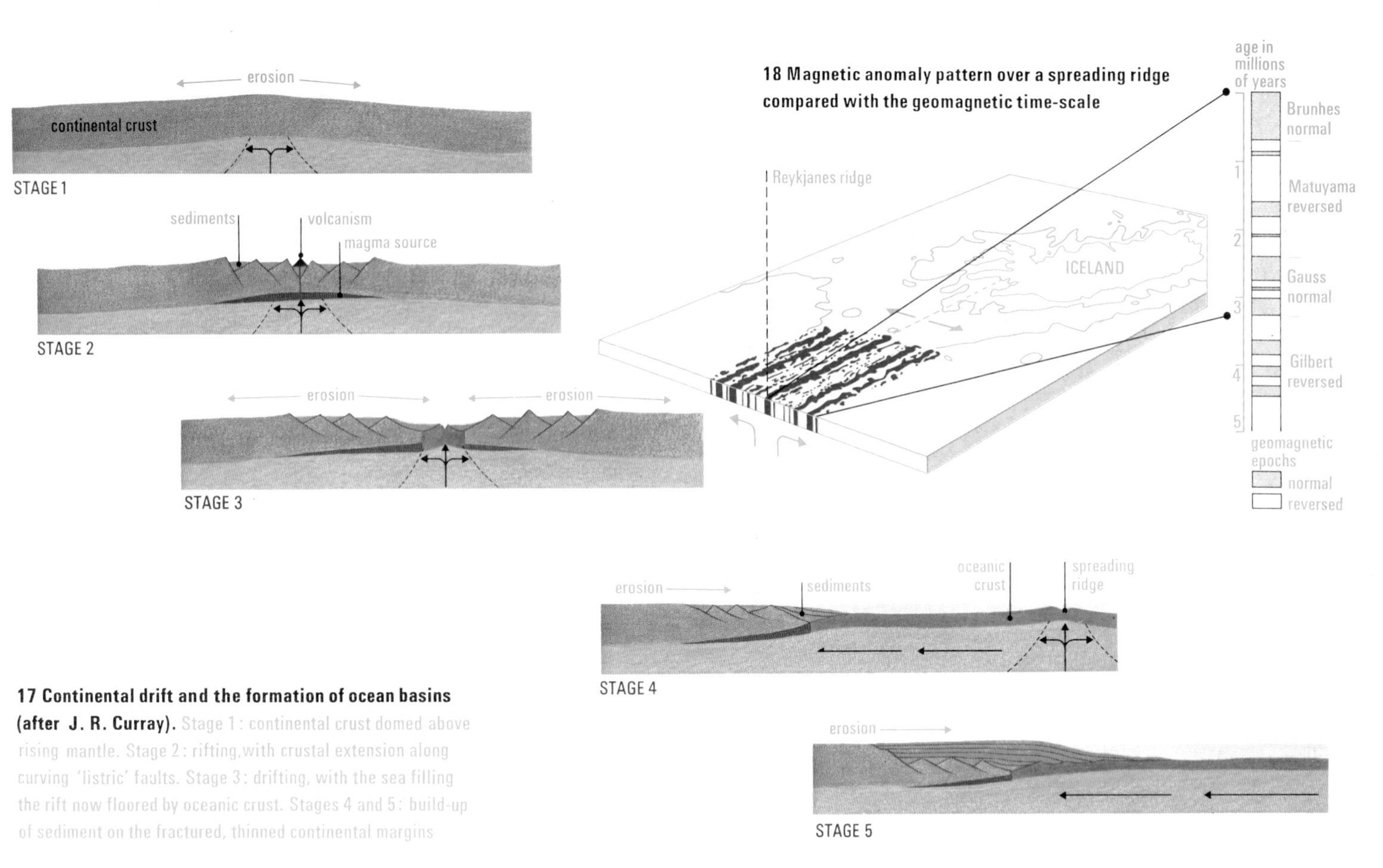

18 Magnetic anomaly pattern over a spreading ridge compared with the geomagnetic time-scale

17 Continental drift and the formation of ocean basins (after J. R. Curray). Stage 1: continental crust domed above rising mantle. Stage 2: rifting, with crustal extension along curving 'listric' faults. Stage 3: drifting, with the sea filling the rift now floored by oceanic crust. Stages 4 and 5: build-up of sediment on the fractured, thinned continental margins

occurring in and around the constantly replenished magma chamber under the mid-ocean ridge. These processes contribute to the mechanism of *sea-floor spreading* in which the ocean floors are continually being added to at the mid-ocean ridges while the older-formed crust moves outwards like a conveyor belt at rates varying from 1 to 8 centimetres per year. Not only is the crust involved in these movements but also the upper mantle to depths of 70–100 km, where the Earth's outer shell or *lithosphere* moves on a semi-molten layer of the upper mantle called the *asthenosphere,* which penetrates to the surface at the mid-ocean ridges. Proof of sea-floor spreading is found in the steadily increasing age of the sedimentary rocks resting directly on the igneous crust encountered in drill-holes at increasing distances from the mid-ocean ridge; it is also proved by the fossil magnetisation patterns disclosed by shipborne magnetometer surveys across the mid-ocean ridges. The patterns are formed when upwelling molten basalt is magnetised as it freezes in the Earth's magnetic field, which we now know reverses in polarity – north becoming south – at intervals of a few hundred thousand years. The crust retains – like a tape recording – a pattern of magnetisation which is identical on either side of the ridge (fig 18) and which can only have been formed by symmetrical outward movement of magnetised crust. All the main ocean basins except the Pacific have formed by the rifting apart of continents when upwelling mantle currents were established under them (fig 17). The Pacific Ocean has clearly decreased in size even though it displays the highest rates of sea-floor spreading: this is because oceanic crust is being consumed on a large scale along its margins.

Early stages in the rifting process are exemplified by the East African Rift Valley (stages 1 and 2 in fig 17) with chains of volcanoes fed from below the crust, and the Red Sea (stage 3 in fig 17). Stages 4 and 5 represent inactive margins characteristic of the Atlantic Ocean.

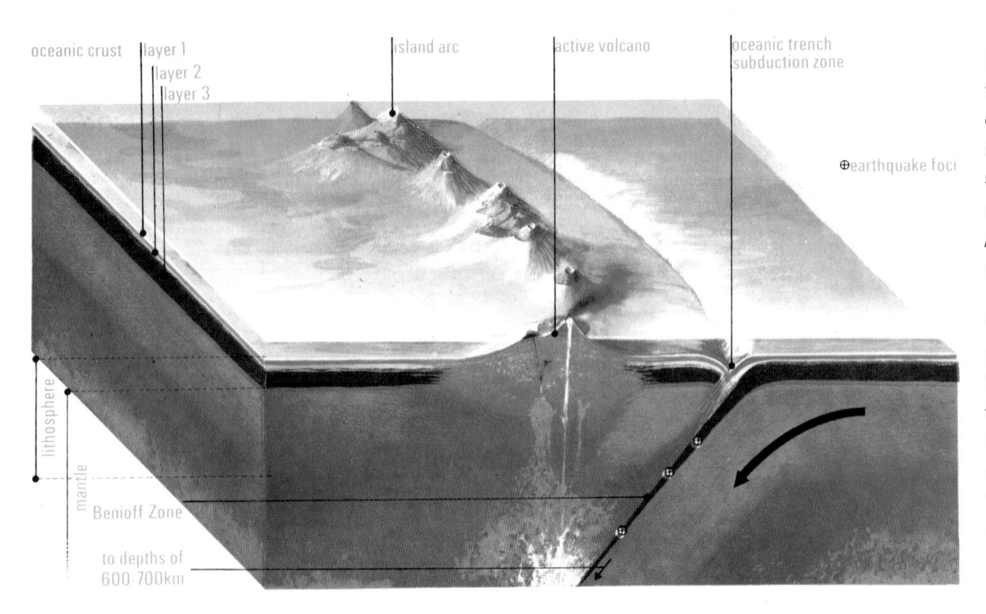

19 Oceanic subduction zone with island arc

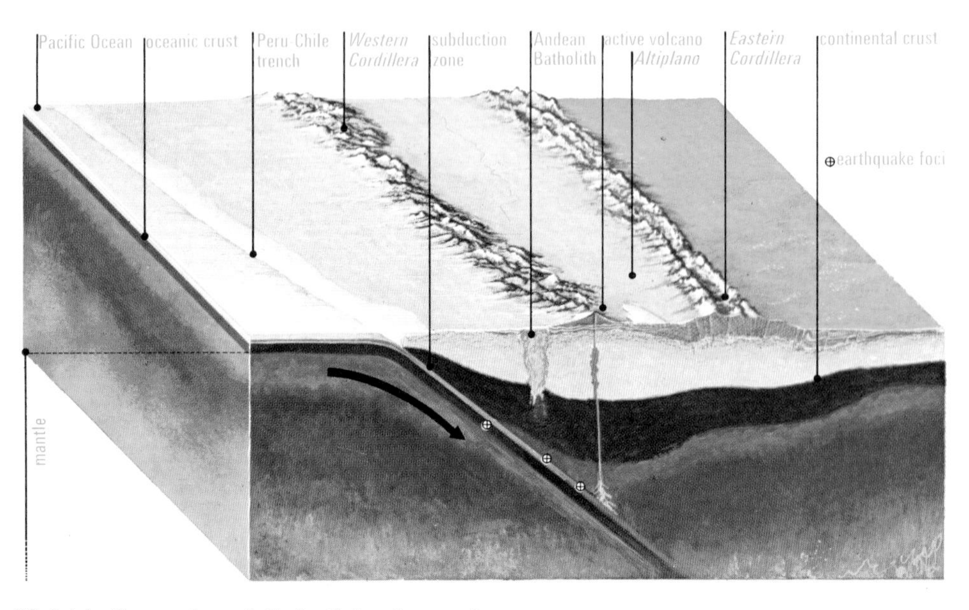

20 Subduction zone beneath the South American continent

Destruction of oceanic crust

In every major ocean and particularly around the Pacific, the oceanic crust is being pushed or drawn down into the mantle as the last surface manifestation of sea-floor spreading. This consumption of oceanic crust takes place in *trenches*, along which two slab-like areas, or *plates*, of the Earth's lithosphere are converging on one another. One of the plates, usually more mobile as a result of active sea-floor spreading from a ridge, is deflected downwards by the other plate; the steeply inclined zone of underthrusting is called a *subduction zone*. As it descends, the underthrusting plate carries oceanic crust back into the mantle, its lower-density surface sediments being mostly scraped off and deformed in the process. All remaining oceanic crust is partially melted or resorbed at depths between 100 and 300 km, but the plate continues downwards to about 700 km before final break-up. Being less dense than the mantle, the molten crustal material rises towards the ocean floor where much of it erupts as lava, and, together with the scraped-off sediments, builds up a chain of volcanic islands, or *island arc*, on the margin of the overriding plate (fig 19). The Aleutian, Kurile, Japanese and Marianas islands form such island arcs. On the west coast of South America, however, where a subduction zone directly borders a continent, spreading oceanic crust is being consumed beneath *continental* lithosphere (fig 20). Here, molten material from the subducted oceanic crust forms a chain of volcanoes which cap a high mountain range, the Andes, into which has been injected a large mass of molten granite (the Andean Batholith).

Frictional effects of movements between adjacent lithosphere plates within subduction zones are indicated by intense earthquake (seismic) activity. The foci of shallow, intermediate and deep earthquakes associated with subduction zones are located on inclined planes ('Benioff zones') which seem to correspond with the upper and lower boundaries of the descending plates.

Active zones and plates It has long been known that earthquakes, volcanoes and high mountain ranges are not randomly distributed over the Earth's surface, but are mostly concentrated in fairly narrow, interconnecting 'active zones'. In more recent years other types of surface feature characterized by crustal movement have been shown to constitute active zones. For example, oceanic ridges and trenches are the sites of vulcanicity and earthquake activity now known to be associated with large scale spreading movements of lithosphere plates. Some of this ridge seismicity has been shown to be caused by movements along the numerous fracture zones which traverse oceanic ridges at right angles to their crests. Many of these so-called *transform faults* are seismically active between those parts of ridges which they offset. Collision movements between lithosphere plates involving continental crust also appear to be responsible for the formation (folding and uplift) of mountain ranges.

The picture of the Earth's outer shell now emerging to scientists shows it to be composed of at least fifteen rigid, virtually undistorted lithosphere plates, seven of which occupy considerable areas (fig 21). Boundaries of plates are of four principal types: *spreading ridges,* where plates are separating and new plate material is added; *subduction zones* (trenches) where plates converge and one plate is consumed; *collision zones,* former subduction zones where continents transported on plates are colliding; and *transform faults,* where two plates are simply gliding past one another, with no addition or destruction of plate material. In summary, almost all the earthquake, volcanic and mountain-building activity which marks the 'active zones' of the Earth's crust closely follows the boundaries of plates and is related to movements between them.

A world-wide system of seismograph stations records the various shock waves from earthquakes. From analysis of their data the relative motions of plates which produced the earthquakes can be ascertained. These plate motions must reflect some form of internal convection within the Earth's mantle.

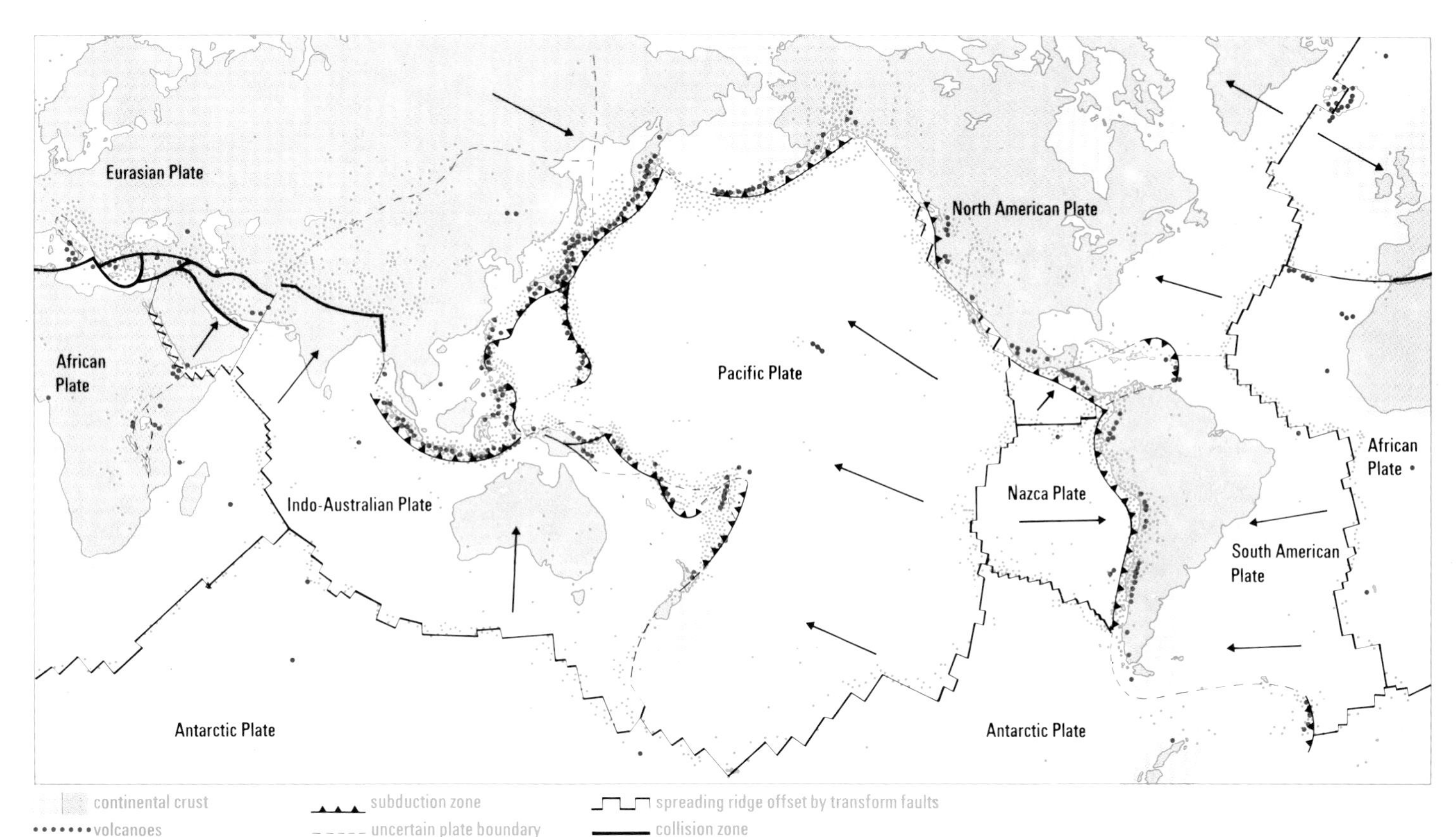

21 Plate boundaries and active zones of the Earth's crust

Plate tectonics is a group of concepts in which the structural complexities of the Earth's crust are ascribed to the interactions of moving lithosphere plates. Within the scene set by the seven major plates, the main rôles are played by: oceanic lithosphere generated at the ridges; continents, microcontinents and island arcs carried as passengers; small ocean basins, such as the Japan Sea, separated from the great ocean basins by island arcs; and inland seas, such as the Black Sea, floored by crust of oceanic type. Between these participants a great many types of interaction are possible; fig 22 shows some of the more common ones. In all cases, consumption of oceanic lithosphere is either actively in progress or has only recently ceased. The simplest situation is shown in (a): oceanic lithosphere is consumed beneath a continent (South America) passively carried on spreading lithosphere. In (b), oceanic lithosphere is consumed beneath an island arc (Japan) separated from the continent (Asia) by a sea-basin with oceanic crust (possibly formed by crustal stretching) carrying a continental remnant (the Yamato Bank). In (c), two island arcs (the Philippine and Marianas arcs) enclose a small ocean basin, while on the opposite side of the main ocean (greatly shortened in the diagram) two major plates (Pacific and North American) slide past each other on a great 'transform fault' (the San Andreas). In (d), a remnant (the Mediterranean) of a much larger ocean (Tethys) is consumed beneath a micro-continent (Turkey) intersected by an earthquake-prone 'transform fault' (the Anatolian Fault) and bounded on the north by partially consumed oceanic lithosphere (the Black Sea) embodied in a major plate (the Eurasian plate). In (e), two plates (Indian and Eurasian) carry continental passengers which have collided with underthrusting to lift up the Himalayas. Other relationships not shown include oceanic lithosphere forced up onto a continent; 'triple junctions' where three plates meet; small spreading ridges located close behind island arcs; and complex systems of island arcs, trenches and transform faults.

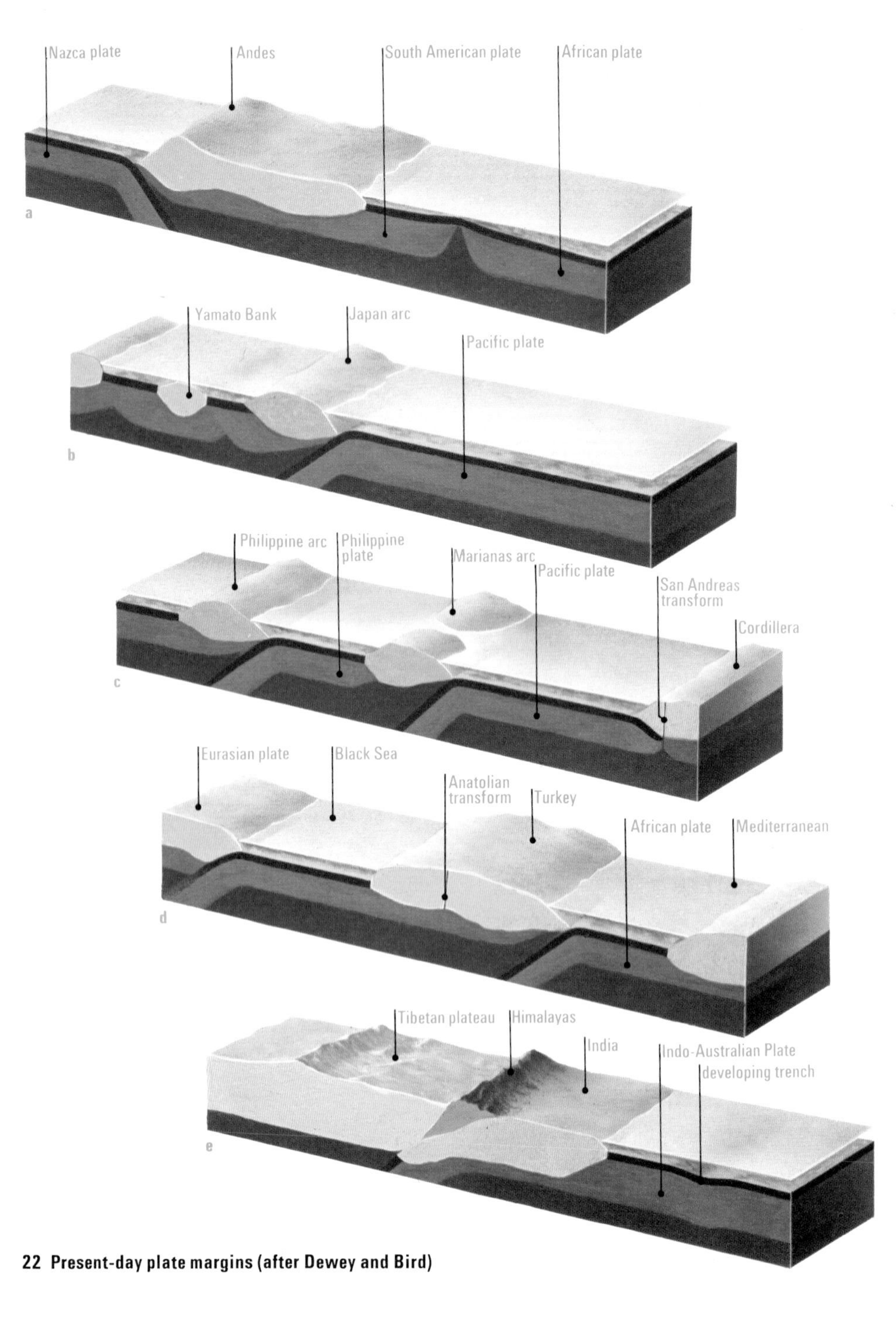

22 Present-day plate margins (after Dewey and Bird)

The continental crust

Three-tenths of the Earth's surface is dry land, and almost all of it lies within the great continental masses: North America, South America, Eurasia, Africa, Australia and Antarctica. These masses are exposures of the distinctive *continental crust,* which also extends beyond the land areas in the continental shelves and under such regions as the Barents Sea.

The thickness and internal structure of the continental crust are known from the study of earthquakes and from seismic echo-sounding using man-made explosions. Fig 23 shows two seismic profiles across Asia produced by these methods. In most places the continental crust consists of a lower, more dense layer and an upper, less dense layer, the junction of which is called the *Conrad discontinuity*. The upper layer, known as the granitic or granitic-metamorphic layer, is composed of 92 per cent igneous and metamorphic rocks, such as granite, schist and gneiss, and 8 per cent sedimentary rocks, such as shale, sandstone and limestone, found mainly in the continental superstructure. It corresponds in average chemical composition to the common igneous rock *granodiorite*.

The lower layer is rather mysterious: nobody really knows for certain what it is made of. At many continental margins, it seems to be continuous with normal oceanic crust and in a few places, such as Lapland and the Southern Alps, where huge dislocations have brought the lower layer to the surface, it is also of oceanic, that is basaltic, composition. But where, as in most places, it is deeply buried, the lower layer could be a greenish-black metamorphic product of basalt known as *amphibolite*. In fig 24, which is based in part on a study of the ruptured and overthrust crust in the Italian Alps north of Turin, the upper granitic-metamorphic layer grades down in composition from average granodiorite to *enderbite*, a high-temperature high-pressure form of granodiorite depleted in radioactive elements. Some geologists think that enderbite and other *granulite-facies* rocks common in regions of ancient crystalline rock compose part or all of the lower layer too.

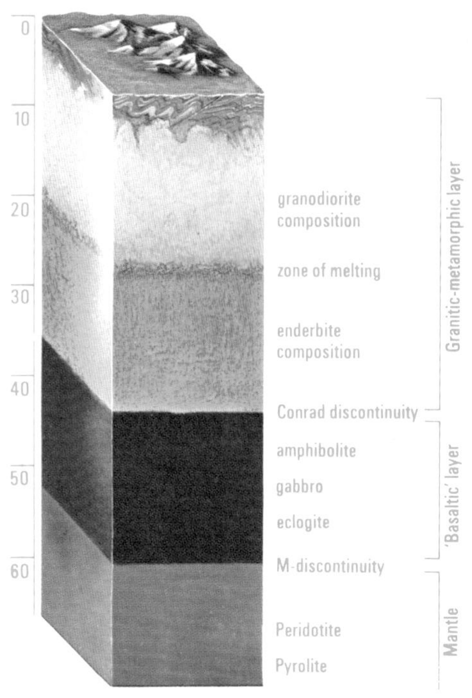

24 Continental crust section

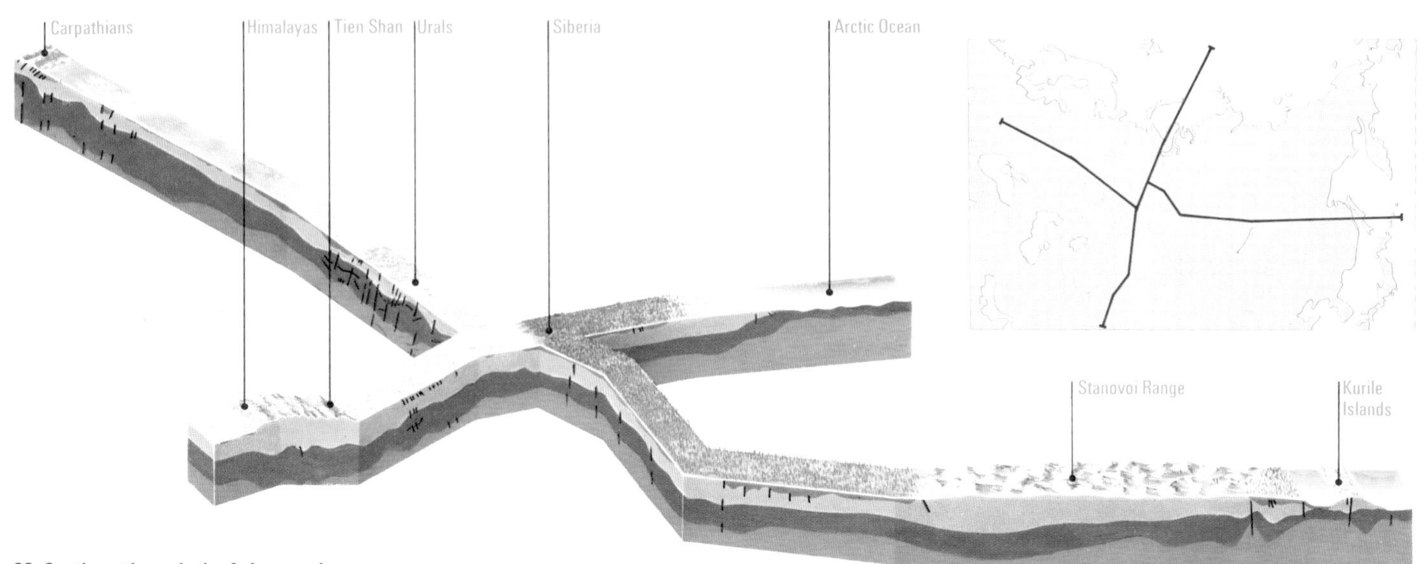

23 Sections through the Asian continent

High mountain chains like the Alps, Himalayas, Andes and Rockies are battlegrounds of crustal deformation and uplift *versus* erosion. Most of them are 'fold-mountains' or *orogenic belts* in which intense deformation has occurred mainly in the past 120 million years followed by rapid uplift in the past 25 million years. Many parts of these ranges are still rising and are active earthquake and volcanic zones. A few high mountain ranges, such as the Tien Shan in central Asia, are uplifted blocks of much older rock. Others are elevated plateaux caught between two branches of an orogenic belt: the Tibetan and Iranian plateaux are examples of such 'median massifs'.

Lesser mountain ranges Moderately elevated, earthquake-free ranges like the Scottish Highlands, the Appalachians, the Urals and the Snowy Mountains of Australia are older fold-mountains, deformed between 250 and 500 million years ago, but uplifted, possibly for the second or third time, within the past 50 million years. Many lesser foothill ranges, like the Jura Mountains, the foothills of the Canadian Rockies and the western Appalachians, are the result of a special type of deformation in which flat-lying sedimentary strata resting on a rigid crystalline basement are rucked up like a carpet on a slippery floor.

Extensive flat regions like the Great Plains of North America, the Russian Steppe, the deserts of North Africa, the Australian outback and the Congo basin are vast structures known as *platforms* and *shields*. Platforms comprise two units: a rigid, stable *basement* composed of old, highly deformed crystalline rocks and a *cover* of younger, flat-lying or gently inclined strata deposited on the planed-off surface of the folded basement. This cover can be very thick. Shields are extensive outcrops of the basement kept clear of cover by persistent uplift of the crust. Many high plateaux and escarpments in Africa and Asia are simply uplifted portions of shields and platforms with, as in the Ethiopian Highlands, extensive volcanic cover.

NORMAN DYHRENFURTH

25 Panorama of the Everest Group

IGS

26 Five Sisters of Kintail from Loch Duich, Ross & Cromarty, Scotland

NATIONAL GEOGRAPHIC SOCIETY

27 Wheatfields in the Great Plains of the USA

The origin of mountain ranges

The powerful deformation and dislocation of the continental crust that occurs along narrow 'mobile belts' subsequently uplifted as mountain ranges is called *orogenesis* – the 'genesis of mountains'. The deformation and uplift affect abnormally thick piles of sedimentary and volcanic rock (up to 20 km thick) accumulated during several hundred million years in deep water along the length of the belt. The whole process of deposition, deformation and uplift is called the *orogenic cycle.*

For many years it was thought that orogenic belts evolved in transcontinental downwarps or 'geosynclines' located over down-sinking currents in the mantle. Deformation occurred when the continental crust beneath the downwarp melted and so lost its rigidity, allowing the sediments above to be squeezed. Uplift of the deformed belt as a mountain range followed when the currents stopped or reversed direction. Now a new theory of mountain-building has been advanced which takes plate tectonics as the causative mechanism. The process starts with rifting of a continent and formation of an ocean by sea-floor spreading (fig 28a). Thick muds and silts accumulate on the continental slope alongside thinner limestones and sands on the continental shelves on both sides of the ocean. After perhaps one or two hundred million years, a trench forms on one side of the ocean and the oceanic crust slides under the continent. In our example, this is followed by consumption of the crust in the open ocean, leading to formation of a volcanic island arc (fig 28b). After further trenches and consumption zones have been established, the island arc is swept into the continent (fig 28c). Finally the ocean closes and the continents collide (fig 28d). The sediments at the continental margins and the volcanic rocks of the island arc are folded and overthrust. Molten igneous rock, generated when the oceanic crust melts in the depths, rises into the crust. Uplift results when buoyant continental crust is thrust under adjacent continental crust.

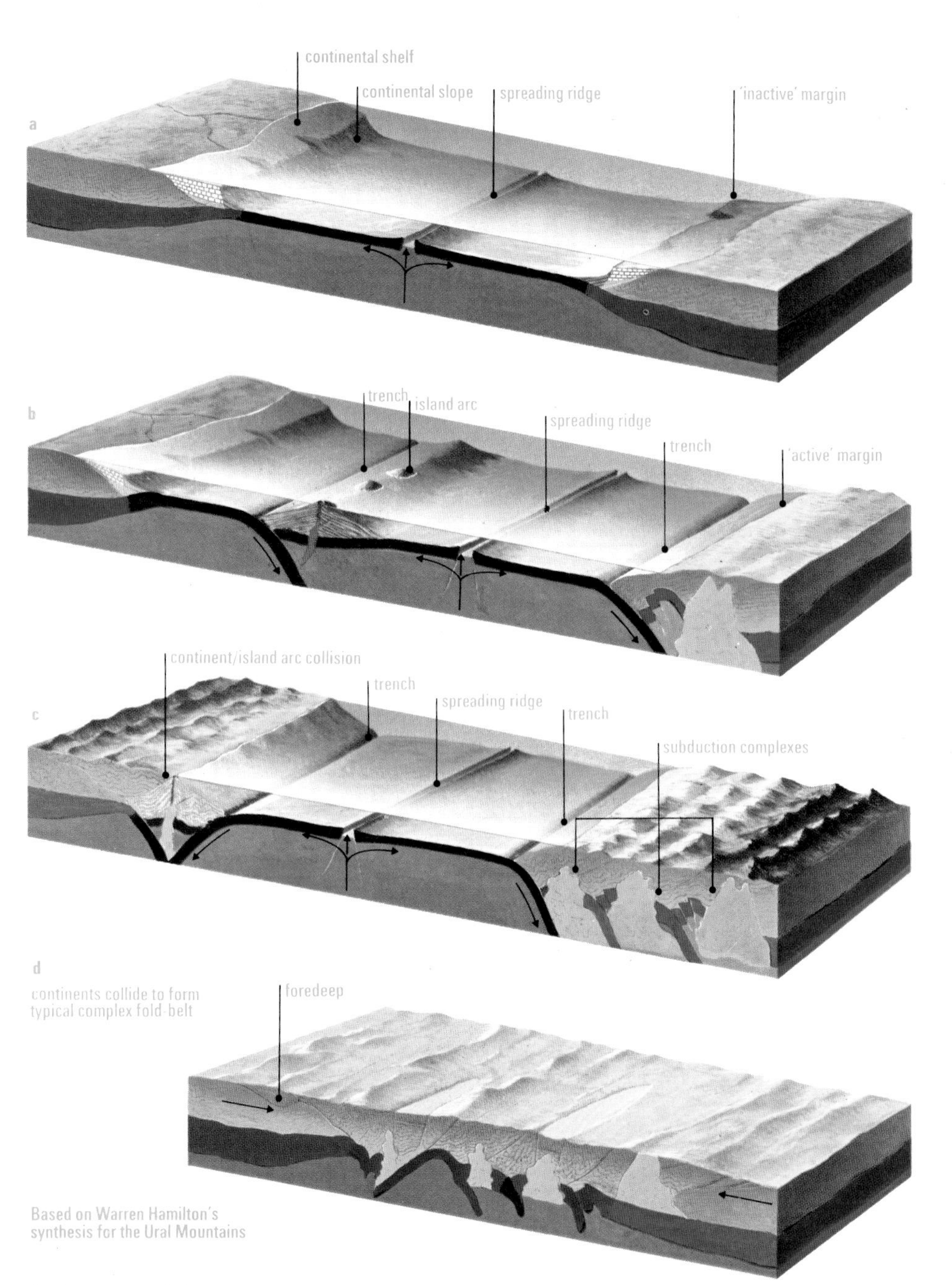

28 Mountain-building by plate tectonics

Structure of the continents

Each continent is a combination of platforms and shields bordered or crossed by younger fold-mountains formed in the past 500 million years. Fig 29 is a section through a typical continent. At least four-fifths of the continental crust, exposed in shields or buried under sedimentary strata in platforms, is Precambrian basement older than 600 million years.

The structure of this old basement is extremely complicated. Fig 30a is a detailed section of highly deformed metamorphic basement 1700 million years old in New South Wales. Steep, 'hairpin' folded structures like this are very common, though in the very oldest crystalline rocks the folds are often near-horizontal as if flattened by the superincumbent load. An association of rocks that characterizes many ancient crystalline terrains in South Africa, India and Australia is shown in fig 30b. Belts of *greenstone* (metamorphosed basalt lava) 'swim' in a sea of granite or metamorphosed granite (granite-gneiss). It is difficult to relate these rocks and structures to those seen in younger fold-mountain ranges. Some geologists maintain that the greenstone belts must be pieces of very ancient oceanic crust, but their location in granite is not easily explained. A curious feature of many old shield areas is that despite the appearance of extremely ancient rocks at the surface, the continental crust beneath maintains a constant thickness.

Large areas of the continents are covered by layers of undeformed sedimentary rock accumulated in shallow seas which once flooded the continental surface. In the centres of some deep sedimentary basins, the basement is depressed to more than 15 km below the surface. The cause of such persistent downwarping, lasting over hundreds of millions of years, is obscure. Most of the sedimentary cover has been deposited in the past 600 million years, but some is very ancient: the famous gold-bearing conglomerates of South Africa are more than 2000 million years old.

In fold-mountain belts younger than 500 million years, the ancient crystalline basement is reactivated and partakes in younger folded and overthrust structures; it commonly builds large parts of the interior regions of younger fold-mountains, such as the Central Himalayas. In fact, in Africa and South America, extensive belts of re-activated crystalline basement are all that remain of some of these mountain ranges.

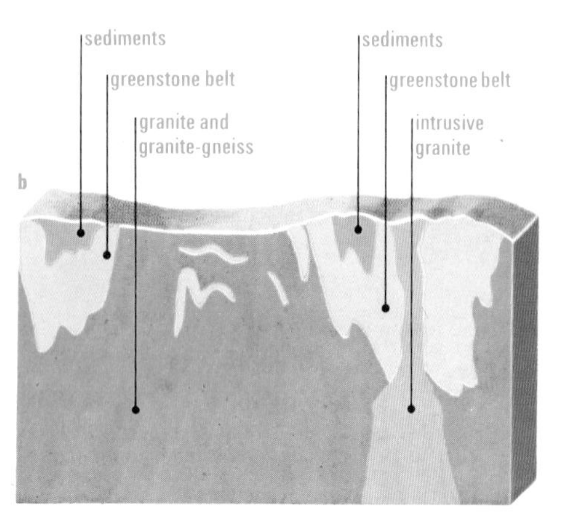

30 (above) Sections of shields

29 (below) Section across a typical continent showing the main structural units

Evolution of the continents

Fig 31 shows the distribution of the main structural units of the continents. Until fairly recently it was not possible to identify older and younger parts of shields with any precision. Now with isotope age determinations, which measure the age of rocks from the amounts of radioactive decay products they contain, the age pattern of a continent can be identified like the rings in a tree-trunk. The continents are seen to consist of ancient cores on to which are welded successively younger extensions. Fig 32 shows growth stages in the exposed and concealed basements of North America and Africa, though it should be emphasized that much of the rock composing younger growth stages is simply re-activated older basement. How has this pattern developed? The plate tectonics theorists imagine a number of small microcontinents' bordered by ocean trenches colliding with each other and with oceanic island arcs to build up larger and larger units. Growth was not regular: world-wide interruptions marked by intense deformation and igneous activity occurred around 2700, 1800 and 1000 million years ago. After each interruption, mountain-building cycles tended to follow parallel, superimposed tracks, in places at right angles to previous patterns, until the next new pattern was established. A surprisingly large proportion of the continental crust was already in existence 2500 million years ago, much of it formed at very high temperatures and pressures rarely attained in mountain belts younger than 1500 million years.

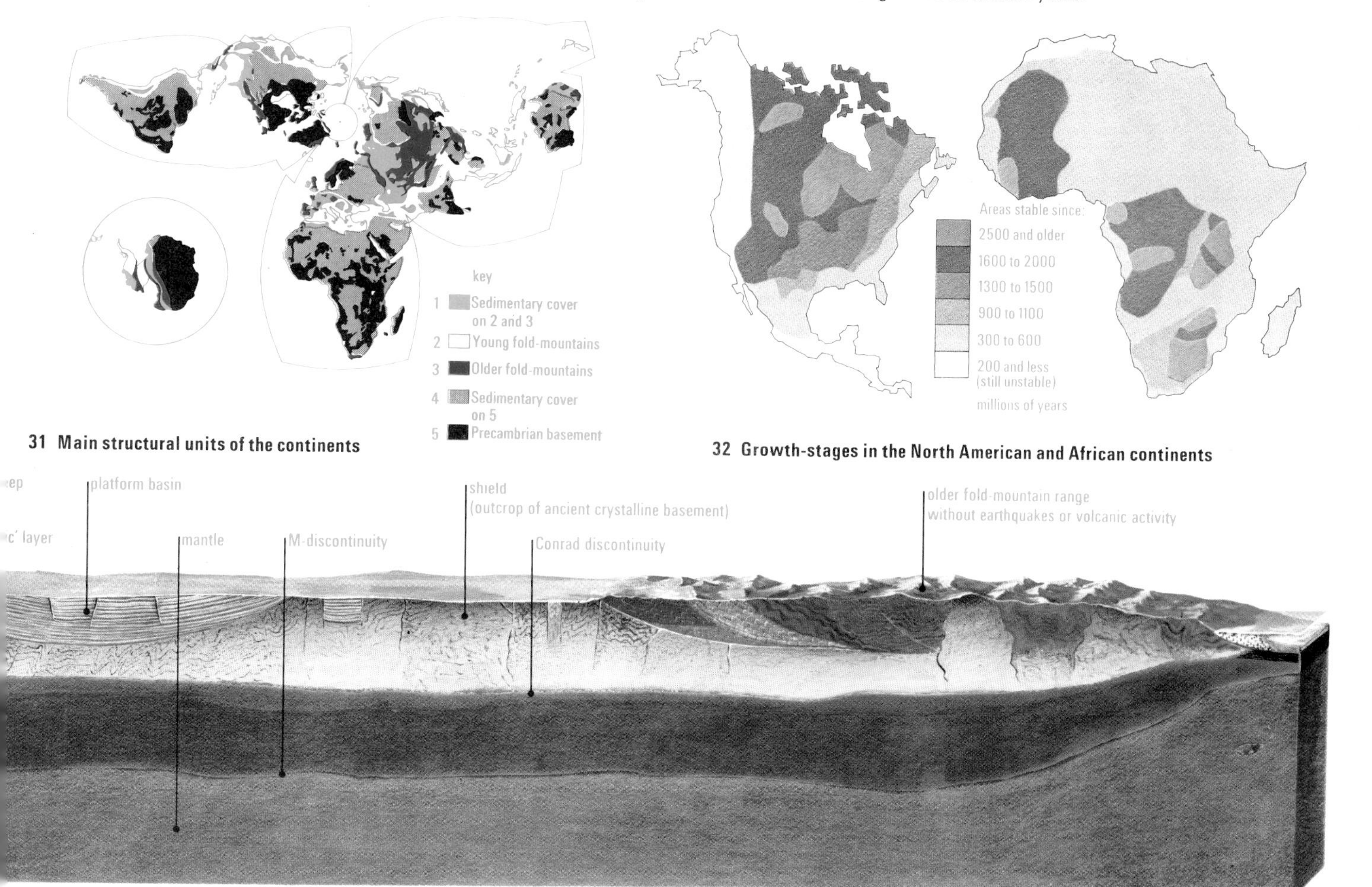

31 Main structural units of the continents

32 Growth-stages in the North American and African continents

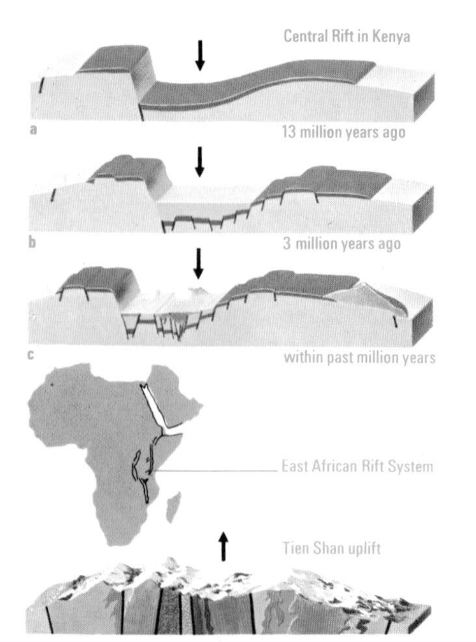

33 East African Rift development and Tien Shan uplift

Vertical movements

Low-lying areas of the continental crust are subject to continuous slow up-and-down movement leading to inundation by shallow seas or to widespread emersion and erosion. The movements are usually differential: one region sinks while the adjacent region rises. They are influenced to some extent by events in adjacent fold-mountain belts but are otherwise independent. Periods of slow uplift and sinking are punctuated by episodes of more rapid movement. This is shown by erosion surfaces, gaps in the deposition of strata and by sudden thickening of rock formations. Rapid movements associated with mountain-building or continental drift cause fracturing of the crust. Elongate blocks of crust subside to form rift-valleys, like the East African Rift System, marked by chains of volcanoes, or rise to form block-mountains like the lofty Tien Shan in central Asia (fig 33).

Young fold-mountains

Though no two young fold-mountain ranges are structurally identical, they have some common features. Their interiors, the *internides*, are metamorphosed and very strongly deformed, usually into great contorted overfolds as in the Alps (fig 34). The rocks involved are deep-water sediments rich in volcanic lavas, perhaps laid down in an ocean. The outer *externides*, built from shallow-water rocks, perhaps deposited on the continental shelf, are strongly folded and overthrust but not metamorphosed. Between the externides and the adjacent platform is a *marginal trough* formed when the main uplift began and filled with detritus eroded from the rising mountain range. The Central Andes lack great folds and overthrusts; they are an older fold-belt uplifted as a block and surmounted by volcanoes. Some young fold-mountain ranges are even now rising at more than a metre per century.

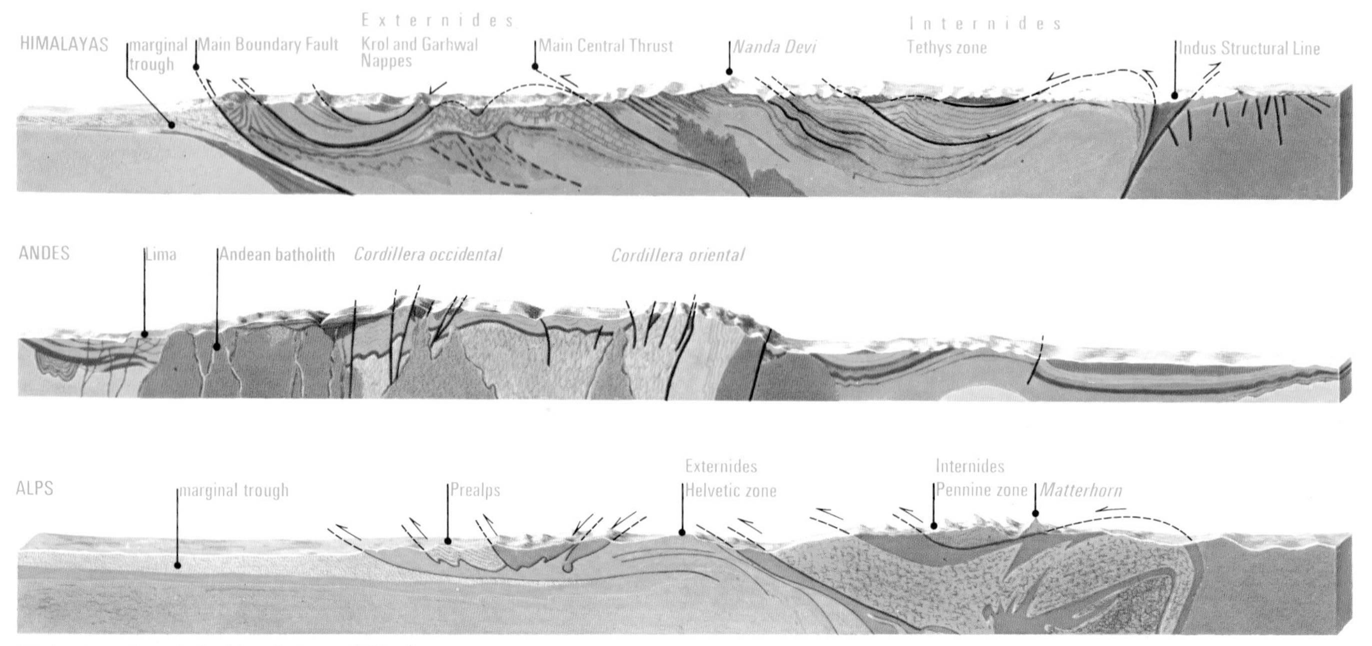

34 Sections through the Alps, Andes and Himalayas

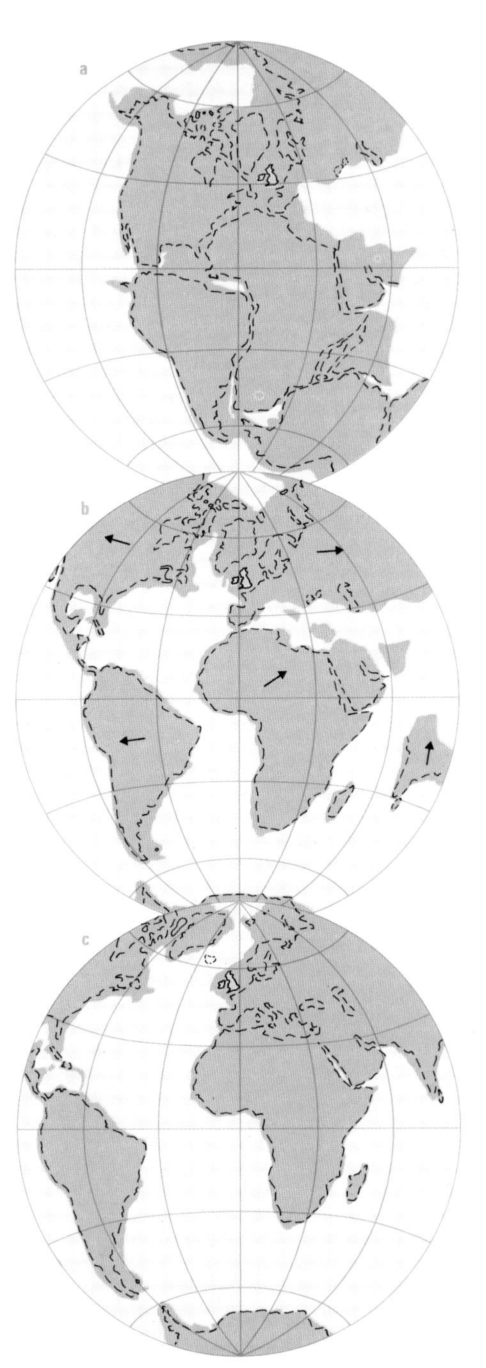

Continental drift

Even before the discovery of sea-floor spreading and the rise of plate tectonic theory, a study of continental geology suggested that the continents we see today are not immovable, but drifting fragments derived from the break-up of an ancient supercontinent, *Pangaea*. The evidence for this came from the way certain continents, now far apart, seemed to fit together, and how geological structures and formations seemed to join up across such a fit; a study of ancient magnetic pole positions measured from rocks in two continents likewise indicated relative movement between the two. The distribution of certain plants and animals in the past, as well as ancient climatic zones, could not be satisfactorily explained unless continental drift was accepted. These considerations suggested that some areas now close together, such as North Africa and southern Europe, had once been far apart; continents now separated by a wide ocean, such as South America and Africa, were at one time united.

Plate tectonics and the recent ocean-floor studies have modified our ideas to some extent and have made the detailed plotting of continental movements possible, but in general they confirm the validity of the classical lines of geological evidence. The pattern of drift is known in detail for the last 200 million years, the period of break-up of *Pangaea;* movements of continents which converged to form *Pangaea* 350 million years ago are known in outline. Precambrian movements are as yet uncharted.

35 The break-up of Pangaea
(a) *Pangaea* 200 million years ago; (b) 60 million years ago: the Atlantic opening and India moving northwards; (c) present day.

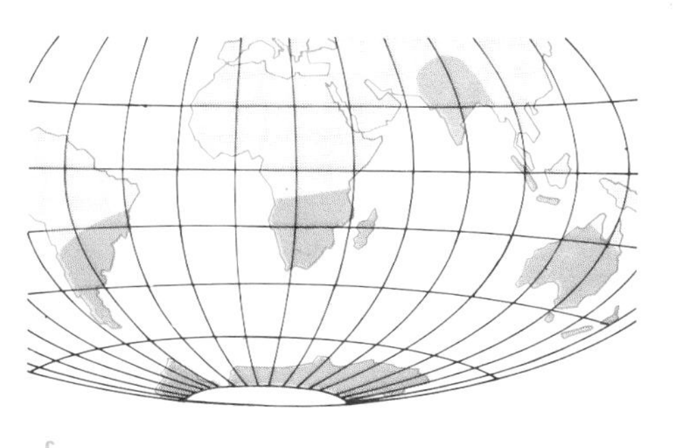

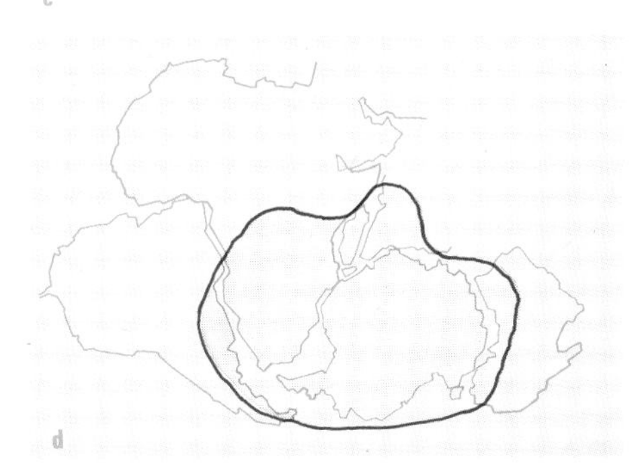

36 (right) The geological evidence for continental drift (a) fit of South America with Africa; (b) separate paths of the wandering pole plotted from rocks in Europe and North America; (c) distribution of fossil leaves of *Glossopteris;* (d) reconstruction of the ice cap which covered *Pangaea* 290 million years ago.

Geological time

Stratigraphy and time The science of stratigraphy aims at reading the story of the Earth in the sequence of layered sedimentary rocks found all over the world. Knowing that in an undisturbed sequence of sedimentary rocks the oldest will lie at the base, the stratigrapher can discover a sequence of events. He uses fossils and distinctive rock layers to compare and correlate sequences from different areas, thus allowing the relative ages of rocks to be determined. The fossiliferous rocks are divided into fifteen *Systems,* and these are subdivided into *Stages* and into *Chronozones.* In parallel with this 'time-stratigraphic' classification is a 'rock-stratigraphic' one based on rock-type, in which *Groups* are divided into *Formations* and *Beds.* Problems can arise, however, when a single rock formation varies in age from place to place and when rocks of the same age contain different fossils because they were formed under different conditions.

Stratigraphic studies yield a geological column showing relative ages of rocks and events. The ages *in years* of some types of rock and the events which have affected them can be calculated from the amounts of radioactive elements and their decay products contained in the minerals composing the rock. Radioactive isotopes of uranium, potassium and rubidium decay at known rates into stable isotopes of lead, argon and strontium respectively. The decay products begin to accumulate in a mineral after it has crystallised and cooled, thus starting a 'radioactive clock'. Isotopic dating is particularly valuable in dividing up the unfossiliferous Precambrian sequences.

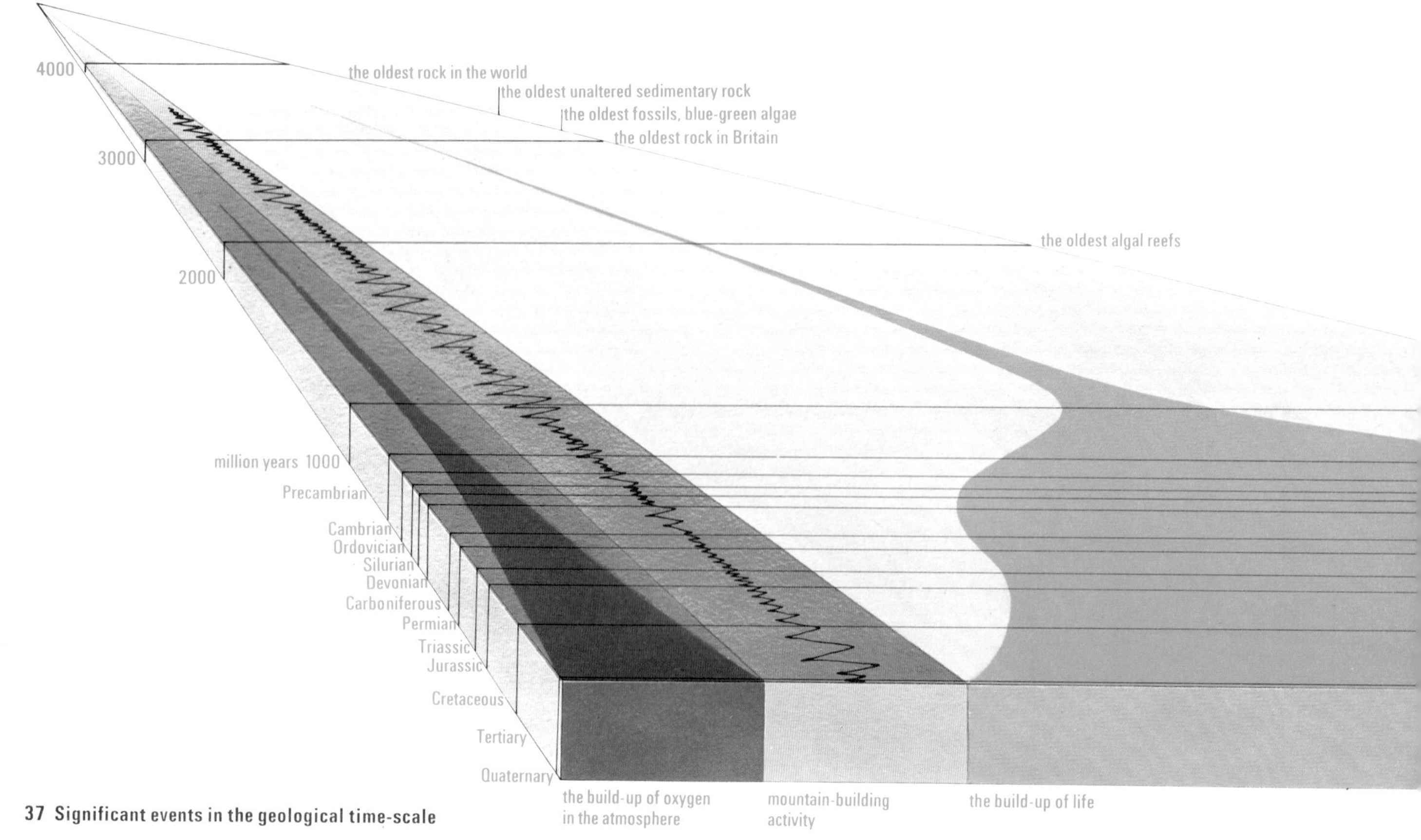

37 Significant events in the geological time-scale

38 **Permian desert in Northern England 250 million years ago**

Ancient climates and geographies Significant changes in climate and geography have occurred even in historical times; over geological time they have been vast and frequent. In the last 600 million years, the British Isles have been the site of shallow seas, mountain ranges, great river deltas, tropical forests, arid deserts and a thick ice sheet. This has come about partly through continental drift but also through actual changes in the average temperature of the planet connected presumably with variations in the Sun's heat output. The evidence for past climate and geography is in the rocks themselves. Examples are red dune-sandstones showing that northern Britain was an arid desert 250 million years ago, and fossil beach deposits 450 million years old in central Wales.

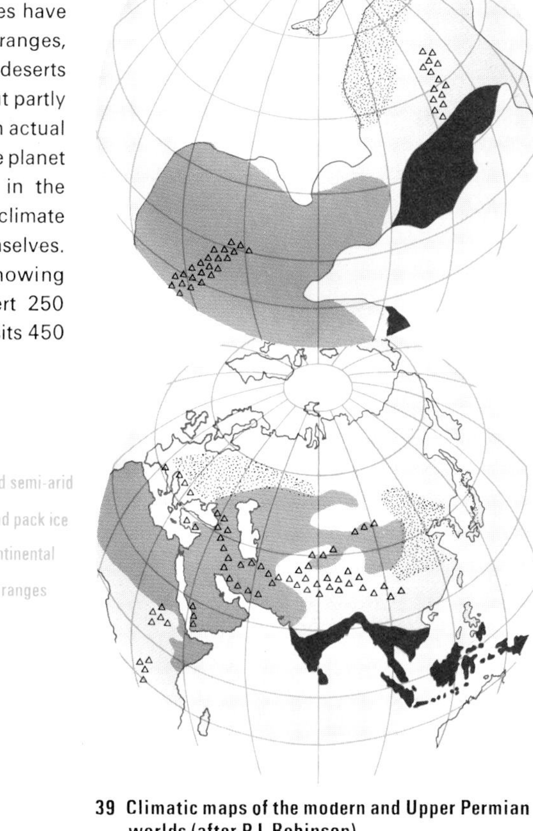

39 **Climatic maps of the modern and Upper Permian worlds (after P L Robinson)**

the first green algae
the first soft-bodied animals
the first shell-fish
the first vertebrates
the first land plants and animals
the age of reptiles
the age of mammals
the age of man

Geological processes

Erosion

All rock at the Earth's surface is disintegrating and decomposing in various processes of *weathering*. Chemical reactions cause the flaking of basalt in fig 42 and frost shattering produced the huge screes in fig 41. Weathering prepares the way for *erosion* which is the wearing away of rock debris and its transportation elsewhere. Together they result in the sculpturing and eventual lowering of the land surface. The downward pull of gravity is the driving force in all forms of erosion, but the main agents by which rock is demolished and removed are rivers, glaciers, waves, currents and wind. Loose material can also fall, slide or creep downhill in the process termed *mass movement*. Many landforms have distinctive shapes which reflect the major processes by which they were moulded. Typical examples include river valleys, glaciated valleys, cliffed coastlines and landslip scars. No agent of erosion works in isolation and all scenery is an assemblage of features shaped in different ways. Fig 43 portrays glaciated landscape in a mountainous environment.

Climatic factors, including the amount and seasonal distribution of rain, snow and evaporation, the range of temperatures and the strength and direction of the wind, control the process of erosion at work in any region. This leads to the development of climatically distinctive landscape types. World climatic patterns are, however, constantly changing. At times during the past million years, ice sheets covered vast areas of today's temperate regions. Changes of atmospheric circulation brought sufficient rain to parts of the Saharan and American deserts to support permanent rivers, whilst some tropical areas, now humid, experienced desert conditions. The processes of erosion in these localities were different from those of today and many features of the present scenery were partly shaped under past conditions. Rocks of varying resistance brought into the zone of weathering and erosion by folding, tilting and faulting are attacked at different rates. Erosion also proceeds rapidly along lines of weakness such as joints and faults. In the arid landscape of fig 40, the erosional forms reflect the differing durabilities of the rock strata and the pattern of

J.A.FREEMAN

40 Valley of the Gods, Utah, USA

jointing, which is also evident in the granite sea-cliffs of fig 45. Over long periods of time, erosion reduces the land to low-lying plains cut across rocks and geological structures of all types. These plains may be raised by crustal uplift to form plateaux. With greater elevation and steeper gradients, rivers cut deep valleys and gorges, as in fig 44. Here the Nile valley is deepened as the Murchison Falls wears back into an ancient erosion surface. As the valleys widen, the plateaux are progressively destroyed, leaving only a series of peaks and ridges at approximately the same height. When uplift occurs before the levelling of a region is complete, partially shaped erosion surfaces form facets of the scenery. These vestiges provide clues to the sequence of events in the evolution of the landscape.

Erosion is rapid in steep areas with heavy precipitation and in semi-arid regions poorly protected by patchy vegetation, but slow in deserts and cold lowlands. The erosion rate for the land area as a whole is estimated as 8.6 cm per 1000 years.

R.A.ANNELLS

41 Screes formed by frost shattering, Iceland

IGS

42 Exfoliation weathering in basalt, N. Ireland

BARNABY'S PICTURE LIBRARY

43 Glacial scenery, Mer de Glace, French Alps

A.C.THACKRAY

44 Murchison Falls, Nile Valley, Uganda

IGS

45 Granite cliffs, Pordennick Point, Cornwall

Sedimentation

The rock materials removed by weathering and erosion are mostly deposited elsewhere as *sediments,* either by settling from suspension or by precipitation from solution. Environments in which sediments accumulate vary widely in their persistence through time, in the size and geography of areas they occupy, and in their climatic conditions. *Sedimentary environments* (fig 46) are essentially *continental* (on land) or *marine* (on the floors of seas and oceans). A transitional *coastal* environment is formed by beaches, lagoons, estuaries and deltas. Sedimentary deposits in continental environments have been laid down either under water or on land. In addition to water-lain muds, sands and gravels of rivers and lakes, there are rock screes, *evaporite* salt deposits and wind-blown *(aeolian)* sands of deserts, and the ice-borne *(glacial)* boulder clays of high mountain and polar regions. Although fragmental *(clastic)* silts, sands and shingles form the bulk of sediments laid down in coastal environments, conditions may permit local accumulation of organic material, such as peat, shell banks and coral reefs.

Marine sedimentary environments range from shallow-water *(neritic)* continental shelf areas, through deeper-water *(bathyal)* areas of the continental slope and rise, to the oceanic *(abyssal)* depths. In general, clastic sediments derived from land grade seawards into finer sands, silts and muds. Accumulations of these sediments on the slope may be dislodged and redeposited by strong gravity currents as 'turbidites' on the abyssal floor. Typical abyssal sediments are slowly accumulating oozes and clays, together with such chemical precipitates as manganese nodules.

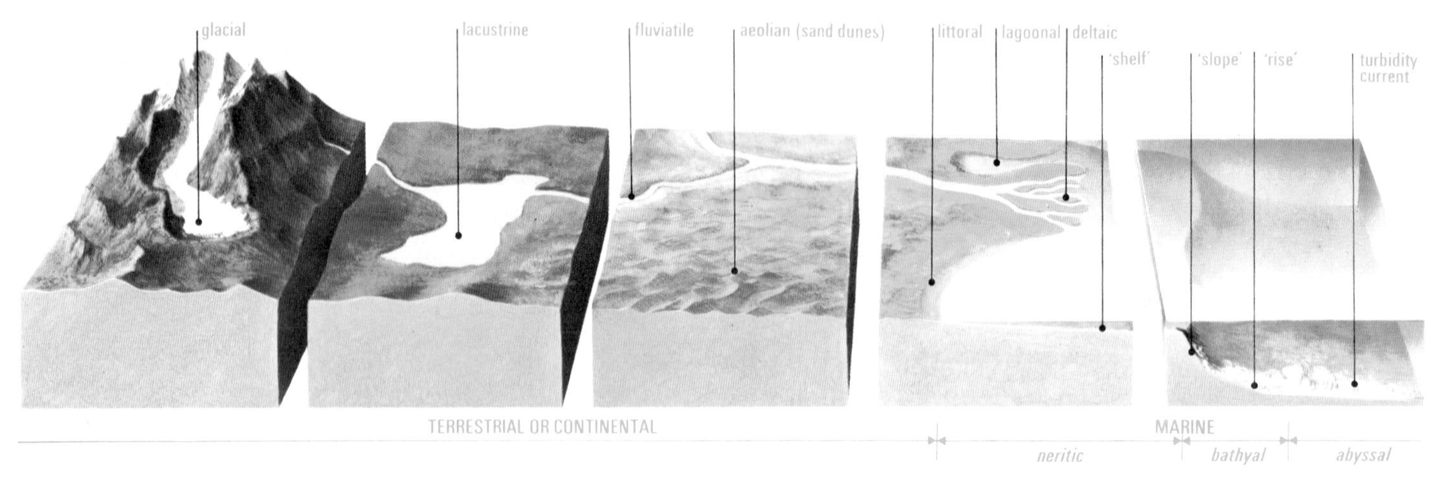

46 The principal sedimentary environments

PICTUREPOINT

47 Desert dunes near El Oued, Algeria

PICTUREPOINT

48 Flood plain of Alligator River, northern Australia

NATIONAL GEOGRAPHIC SOCIETY

49 Coral reefs off the Honduras coast

Sedimentary rocks Covering more than two-thirds of the Earth's surface, sedimentary rocks and sediments provide the main record of past geographies, shifting coastlines and changing climates. Because fossils are restricted to them, they yield much information on the changing pattern of plant and animal life, and have supplied the principal means of establishing the divisions of geological time.

The distinction between sediments and sedimentary rocks is rather arbitrary, but is generally based on a change to a more compact and hardened physical state. Thus, gravels are naturally cemented into *conglomerates*, rock screes into *breccias;* sands become indurated to *sandstones;* muds are transformed into *mudstones* and fissile *shales;* limy deposits gradually harden to *limestones;* peat ultimately becomes *coal.* The chemical and physical changes which lead to rock hardening are termed 'lithification'; cementation by mineral matter deposited from groundwater, pressure due to burial, and stress in mountain-building are all factors.

Sedimentary rocks are grouped into three major categories: (a) *fragmental* (or clastic), by far the largest of the groups, with many variations in grain size and composition; (b) *chemically precipitated,* subdivided according to their chemical composition; and (c) *organic*, made from the fossil remains of animals or plants. The commonest fragmental rocks are *rudaceous* ('pebbly') breccias and conglomerates; *arenaceous* ('sandy') grits, sandstones and siltstones; and *argillaceous* ('muddy') clays, mudstones and shales. Chemically precipitated rocks include *evaporites* (like rock-salt and gypsum), some varieties of ironstone, siliceous cherts and many different limestones. Coal and richly fossiliferous limestones are examples of organic sedimentary rocks.

Most sequences of consolidated sedimentary rock show distinct layering, known as *bedding* or *stratification*, caused by slight alternations in composition or by changes in rock-type. Various styles of stratification, coupled with other bedding-plane features, including fossil rainpits, suncracks, scour and ripple marks, provide many clues to the nature of ancient sedimentary environments. Ancient patterns of wind and water flow can, for example, be deduced from such structures.

50 Sea-cliff in well-bedded sedimentary strata (Middle Old Red Sandstone), Mainland, Orkney Islands

Deformation

All sediments and most volcanic rocks are horizontal when first laid down. In stable regions of the continental crust and over much of the ocean floor, they have stayed horizontal for many millions of years. But in unstable regions of the crust, the strata have been folded, squeezed and fractured, processes collectively called *deformation*. The folds can be all sizes from huge arches or overfolds scores of kilometres across to microscopic crinkles. Fig 51 shows an average-size fold in Carboniferous strata in Cornwall. Its centre-line is horizontal; the fold is described as *recumbent*. Fractures can be anything from the great dislocations which bound the plates of the Earth's crust, like the San Andreas Fault in California, to minute displacements in single crystals. The way in which rocks deform under stress is determined by their composition, the temperature and pressure under which they are confined, the pressure of water contained in pores in the rock and the *strain-rate*, that is the amount by which the rock is compressed in a given time. Depending on these factors, the rock may fracture, crush, fold or flow plastically. For example, a hard rock – quartzite – compressed rapidly at low, near-surface temperatures and pressures will fracture or even be crushed to powder (which on hardening forms a rock called *mylonite)*. The same quartzite, deep in the interior of an active fold-mountain range, at high temperatures and pressures, under steady stress, will fold into intricate shapes without cracking. The quartz grains in the rock adapt by sliding over each other and by dissolving at stress points. In this way, rigid crystalline basement rocks become as ductile as the sediments which rest on them: thus were formed the gigantic convolutions of the central Alps. Structures of this type contrast strongly with the brittle-style fracturing and simple concentric folding seen in rocks deformed at shallow depths.

51 Recumbent fold in Carboniferous strata, St Gennys, Cornwall

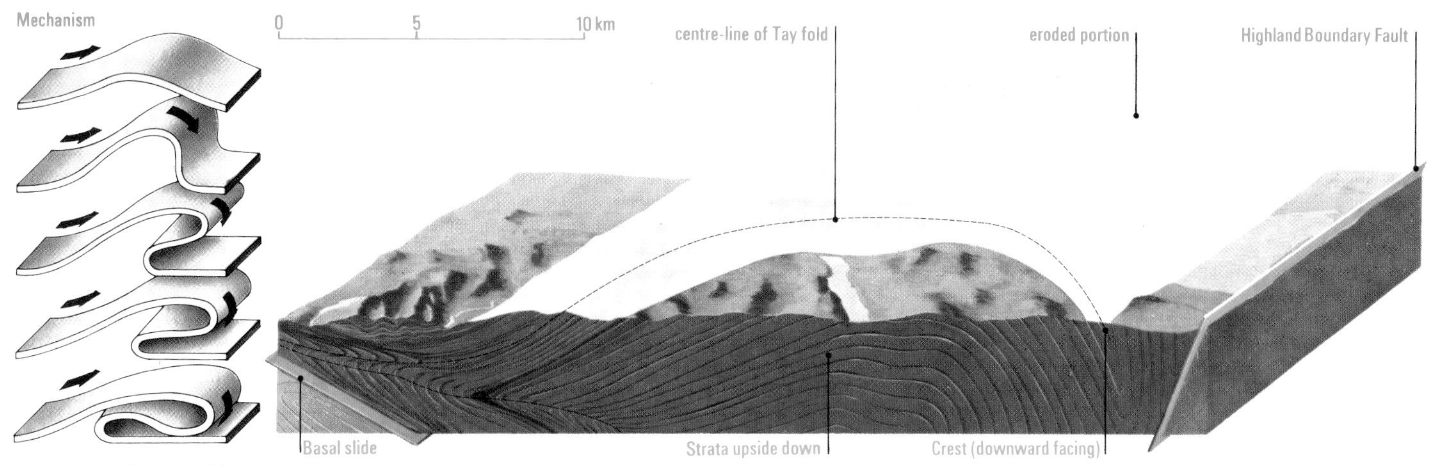

52 The Tay Nappe and its emplacement

J.G.RAMSAY

53 Overthrust Permian rocks of Glarus Nappe, Segnes Pass, Switzerland

J.G.RAMSAY

54 Re-folded fold in gneiss, Pennine Alps, Switzerland

Nappes (French : a sheet) are very large flat-lying masses of rock which have been driven over geologically younger rocks for very long distances. The name is also applied to huge recumbent overfolds like the Tay Nappe in the Grampian Highlands, shown in fig 52. Note that the country beneath the fold is made of upside-down strata. Most nappes have travelled on a flattish thrust-plane and consist of the right-way-up part of an overfold, the inverted lower part having been eliminated by extreme stretching. Fig 53 shows a mountaintop in the Alps composed of dark continental sediments 250 million years old, part of the Glarus Nappe, thrust over much younger marine sediments belonging to the inverted limb of the Glarus Nappe. Some nappes can be traced back to a 'zone of roots' where the rock layers are seen to plunge steeply underground. Nappes are especially characteristic of mountain ranges like the Alps and Himalayas formed by the collision of continents.

Repeated folding The rocks in most fold-mountains have been folded more than once. This is especially evident in ancient folded basement rocks caught up in younger deformations, but repeated folding is also common in strata deformed for the first time, one episode of folding apparently following rapidly on another. Repeated folding is seen on all scales, though it is not usually possible to see more than two foldings in a single outcrop. Fig 54 is a very clear example from the Alps : the crest of a tight fold in banded gneiss is sharply bent over by a secondary fold. The same thing is seen on a large scale in maps and aerial photographs of fold-mountains, where repeated folding causes sharp changes of direction of rock layers. An extreme example of repeated folding is seen in very old basement rocks such as the 3000 million year old Lewisian Gneiss in northwest Scotland in which six successive deformations can be identified. The same rocks were again folded four times when the Scottish Highlands were formed 500 million years ago !

Faults and Earthquakes

Almost all rocks except plastic clays are traversed by narrow cracks or *joints* caused by shrinkage or tensional forces. Fractures across which there is measurable displacement are called *faults*. Fig 55a shows the three main types of movement on faults. Fig 55b shows some very large faults in Britain. The largest and deepest, called *geofractures*, have long histories of movement, at various times behaving as normal, reverse and strike-slip faults.

An earthquake is a violent shaking of the ground, visible and wave-like in severe shocks, normally lasting for a minute or less. The record is seven minutes in the great Alaskan earthquake of 1964, which started without warning as a gentle rolling motion lasting for 40 seconds followed immediately by about four minutes of hard shaking, which then gradually subsided. Severe aftershocks went on for several hours, followed by nine months of lesser shocks. Most earthquakes are associated with large faults. Fig 56 shows a popular theory of the sequence of events. The steady movement of one block of crust sliding past or over another has, since the last earthquake, been arrested: the two sides of the fault have stuck together. Stress builds up and the rocks bend like a spring. Suddenly the stress exceeds the strength of the bond and the blocks move. Stress is relieved and the rocks spring back to shape. Rapid vibrations travel through the rocks from the spot or *focus* where unsticking occurred. The point on the surface vertically above the focus is called the *epicentre*. The magnitude scale most commonly quoted in news reports is the *Richter Scale,* which is based on the amplitude of ground motion, zero being the smallest recorded earthquake and 8.9 the largest. Many earthquakes originate at plate margins, where plates move past each other on vertical 'transform faults' like the San Andreas, or which dip down hundreds of kilometres beneath the continental margin, as in Alaska. Shallow-focus earthquakes originating no deeper than 70 km are the commonest and most

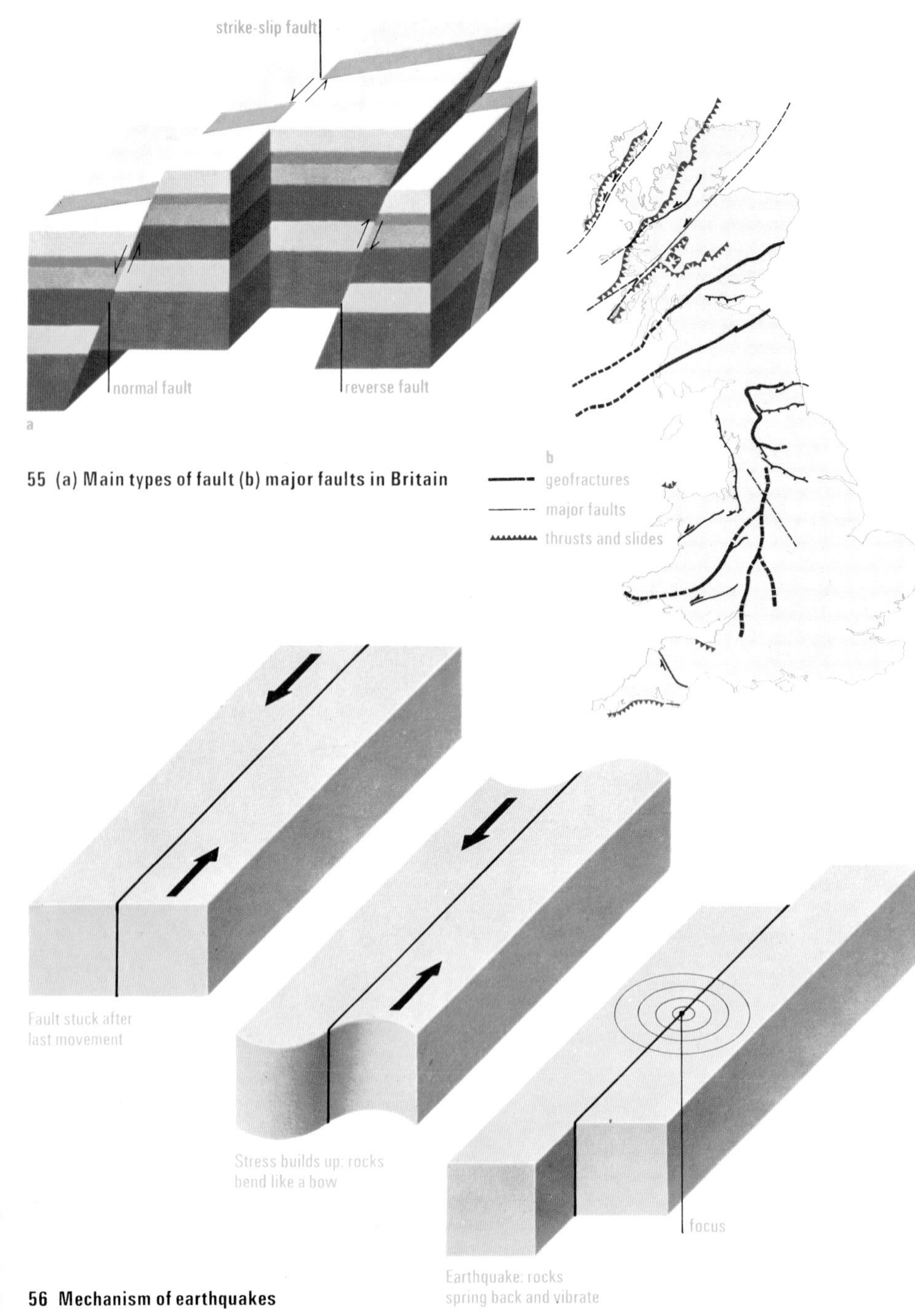

55 (a) Main types of fault (b) major faults in Britain

56 Mechanism of earthquakes

destructive; the worst damage is to buildings sited on soft alluvium.

The most dangerous side-effects of earthquakes are avalanches and seismic sea-waves (called *tsunamis*). One particularly disastrous avalanche occurred in the Peruvian earthquake of 1970. The snow and ice cornice of Nevados Huascarán, the highest peak in Peru, broke off, fell 1000 metres, picked up boulders and other rock debris and then hurtled at 480 km per hour down the Santa Valley, careering up the valley sides as it banked the corners, to demolish the village of Yungay. Tsunamis are caused by rapid uplift or depression of the sea floor during an earthquake. They travel at around 750 km per hour, but in the open ocean the 800 km-long waves, only a metre or two high and taking a whole hour to pass, are too insignificant to observe. But when they run on to a shelving shore they rapidly rise to heights of 12 metres or more, causing massive destruction.

57 Peru earthquake: Huascarán and Yungay before

ZFA

58 Peru earthquake: Huascarán and Yungay after

US GEOLOGICAL SURVEY

59 Tsunami damage, Kodiak, Alaska

R.E.WALLACE, US GEOLOGICAL SURVEY

60 Stream valley offset by the San Andreas Fault

US GEOLOGICAL SURVEY

61 Elementary school damaged in the Great Alaska Earthquake, 1964

Metamorphism

In certain situations, natural heat and pressure can cause pre-existing rocks to crystallize or re-crystallize. Their constituent elements either re-group in new minerals or re-form in larger crystals of the original minerals. This is *metamorphism* and the type of metamorphic rock produced depends on the composition of the original rock, the temperature and pressure, and the amount of water present or introduced. There are two main kinds of metamorphism: regional and contact. *Contact metamorphism* occurs around molten igneous bodies injected into cool country rocks, and the zone of alteration in the country rock is called a *contact aureole* (fig 63). The size of the contact aureole and the degree of metamorphism depend on the girth of the igneous body and the temperature at the contact. The best-known products are *hornfels* formed from rocks with sizeable clay content and *marble* formed by recrystallization of limestone. *Regional metamorphism* occurs in the interiors of fold-mountains where the heat-flow is high over a large region; it is closely linked with deformation. A heat-gradient is established which determines the distribution of different kinds of metamorphic rock – high-grade in the centre, low-grade on the outside. Fig 62 is a map of Scotland corrected for subsequent fault displacements, showing the pattern of thermal grading established 500 million years ago. One of the commonest products of low-grade regional metamorphism is *slate,* which cleaves along planes of tiny mica flakes which grew in the original mudstone at right angles to the direction of principal stress. At higher grades, the mica flakes become visible to the naked eye in *mica-schist.* Finally, before the rock melts, the dark and pale minerals segregate to give a banded rock called *gneiss.* The grade of metamorphism is shown by *index* minerals (fig 62, key) which appear under specific conditions of temperature and pressure. High-grade metamorphic rocks can be downgraded if metamorphosed again under low-grade conditions.

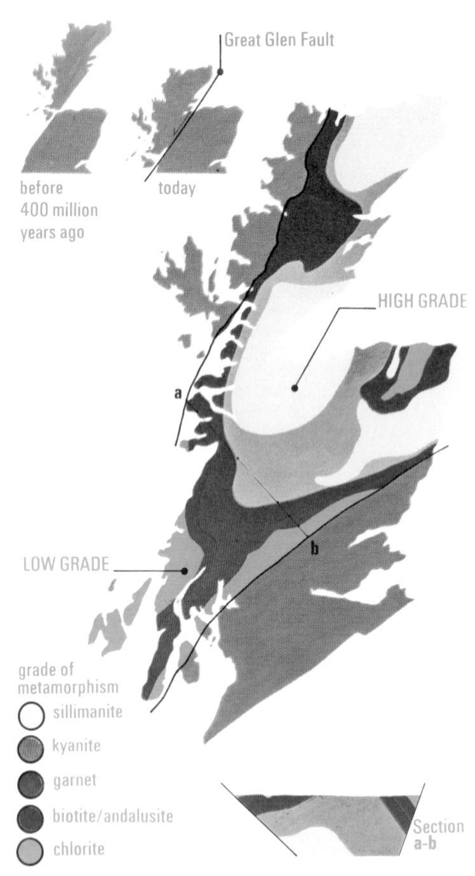

62 Regional metamorphism of Scottish Highlands

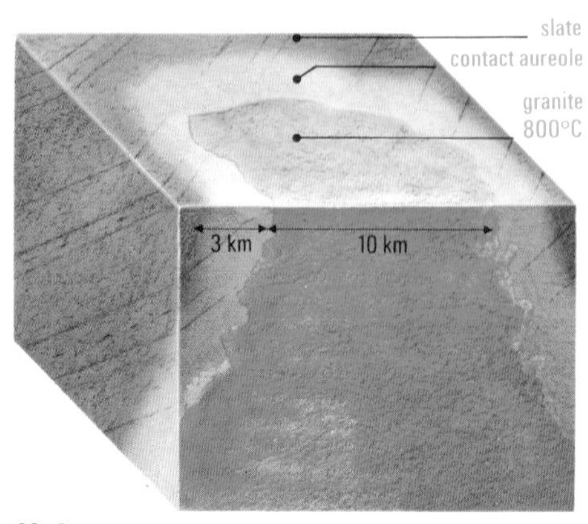

63 Contact aureole

Magma

Molten or semi-molten rock which originates underground is called *magma*. When it cools it forms *igneous rocks*. Cooling may take place underground, when the products are described as *plutonic* or *intrusive*, or at the surface to give *volcanic* or *extrusive* rocks. Magmas are composed of silicates of various metallic elements, chiefly calcium, sodium, potassium, aluminium, iron and magnesium, with varying amounts of dissolved gases and water. Contrary to popular notion, the rocks in the Earth's interior, though very hot, are not normally molten. The pressure of overlying rock prevents melting, which only occurs when the pressure is relieved or if the rock composition is modified by the addition of water or other fluxes. Magma, once formed, tends to migrate towards the surface, either following fractures or drilling its way up by a combination of chemical attack and thermal shock in the manner of a thermic lance. Rapidly chilled magmas form glassy rocks. Slower cooling allows crystals to form; and the slower the cooling the larger the crystals. Some magmas crystallize to rocks composed of one mineral only, but most igneous rocks comprise several different minerals, some of which crystallized before others (fig 64a).

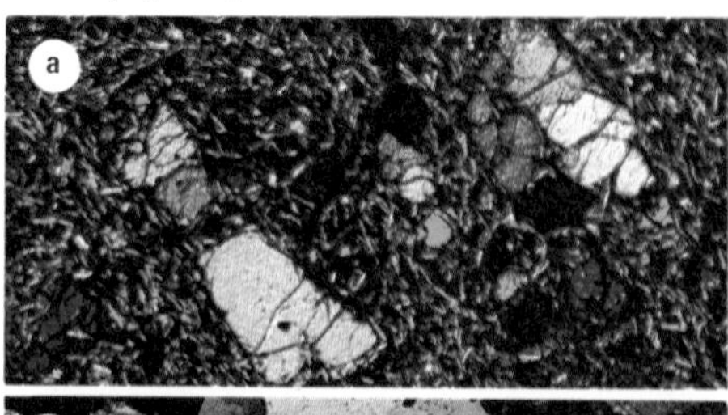

64 Thin sections of (a) olivine-basalt (b) granite

Formation and emplacement of igneous rocks The number of different kinds of igneous rock runs to many hundreds even though the variety of rock-forming minerals is relatively small. Minerals 'freeze out' of cooling magma, each kind crystallizing at different temperature so that, in most instances, the magma has no sharp freezing-point. Many minerals become unstable when they cool in the magma and react chemically with the surrounding liquid to form new minerals stable at the lower temperatures. Sudden cooling of magma can freeze-in and preserve higher-temperature minerals like olivine (fig 64a: olivine-basalt). Otherwise, such early-formed crystals tend to settle like snow towards the bottom of a mass of slowly cooling magma, thus producing *differentiated* magmas of modified composition. These crystal separations, over prolonged periods, give rise to many different kinds of igneous rock. Within, for example, massive flat-lying bodies of basaltic magma called *lopoliths* are formed thick layers of differentiated rock ranging in composition from dark olivine-rich peridotite to pink granite. Various magmas, differentiated after resting in pockets – *magma chambers* – high in the crust, are intruded into steep fractures as thin sheets called *dykes* from which magma may spread laterally along the strata to form flat-lying *sills* and blister-like *laccoliths*. Magma which reaches the surface erupts in volcanoes. *Basaltic* magma, derived from partial melting of restricted areas of the mantle, is the parent of most igneous rock varieties formed beneath the oceans; it also feeds volcanoes in continental rift-valleys. Many island-arc and mountain-chain volcanoes erupt basaltic rocks, while others continually erupt lavas of *andesitic* and *granitic* composition. Andesite is produced where oceanic crust descends below these active areas and undergoes selective melting. Differentiation of this andesite magma and also deep melting of the continental crust give rise to granitic magma. This, in time, rises as enormous red-hot, gas-rich blobs of semi-molten rock which ultimately gather higher in the crust to form vast *batholiths* of granite and related rocks (figs 64b, 65).

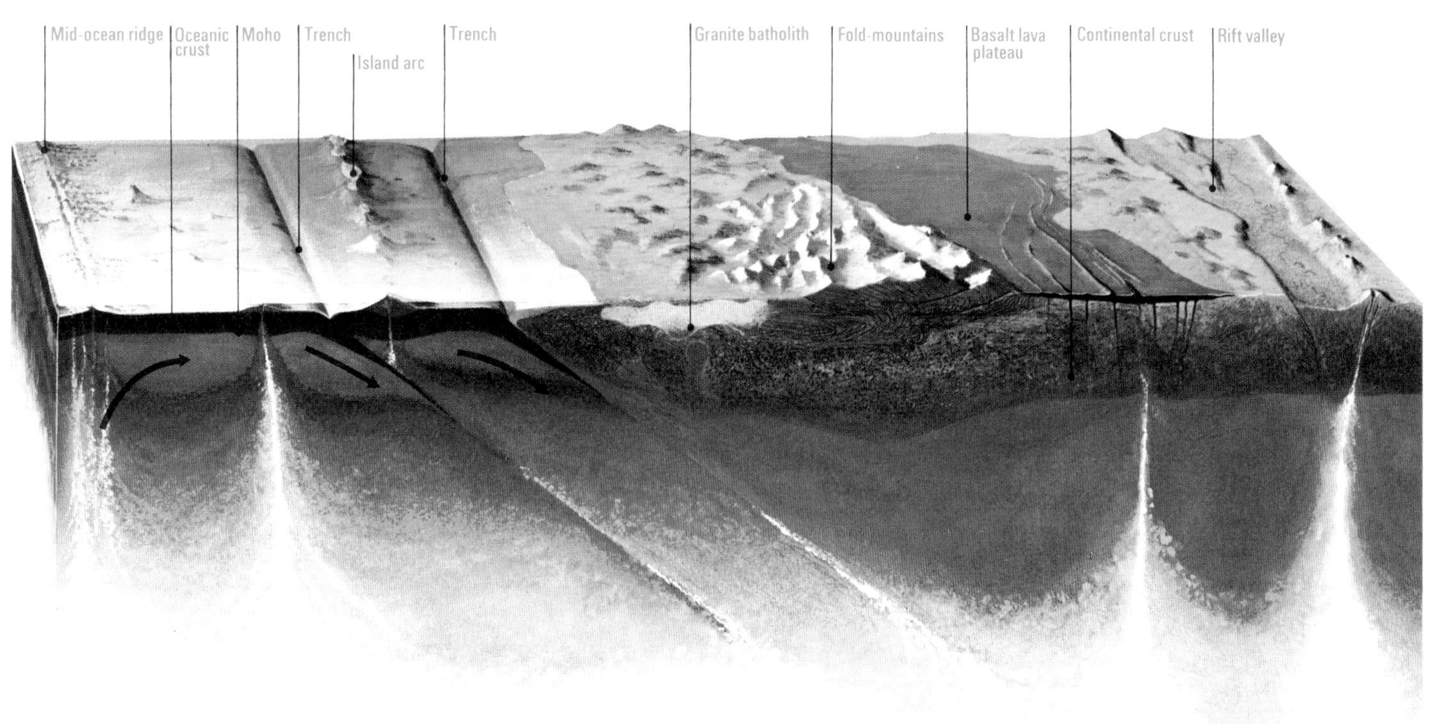

65 Igneous activity

Volcanoes

A volcano forms where magma pours out at the surface of the Earth. Usually, but not always, the volcano is a solitary mountain. Magma under pressure inside the Earth contains varying amounts of dissolved gas. Magma that reaches the surface is called *lava.* During eruption, the gas separates from the lava. When lava is thin and runny, gas escapes easily and the eruption is mild. But if lava is thick and pasty, the trapped gas explodes out, shattering the lava to fragments. The fragments are all sizes from large 'bombs' to fine-grained dust or ash. Volcanoes which erupt explosive pasty lava are usually found in island arcs and young mountain chains, particularly where oceanic crust is sliding under a continental margin. Volcanoes with runny basalt lava are chiefly found in areas where the Earth's crust is splitting apart. Thus many (though not all) basaltic volcanoes are found in the oceans. Many submarine eruptions must go unnoticed because they take place deep beneath the ocean. Eventually a submarine volcano may grow so large that an island is formed. Surtsey is a new island off Iceland which grew in this way in 1963.

Fissure volcanoes build up over long cracks in the crust. Commonly the lava is runny basalt which spreads out as thin flows to form plateaux like those in Iceland today. These are called *flood basalt* or *plateau basalt* eruptions. A *central volcano* is fed by a single pipe and builds a cone with a summit crater. Between eruptions the crater may fill with rainwater forming a crater lake. Basalt eruptions from central volcanoes produce thin lava flows, spreading over a wide area to form flattish volcanoes *(shield volcanoes)* shaped like upturned saucers. *Strato-volcanoes* (fig 71) are characterised by steeper slopes and less voluminous eruptions in which short lava flows pile up around a central crater. Many volcanoes have an eruptive cycle where initially gas pressure is high and activity explosive. As pressure is released by explosions the style of activity becomes milder, terminating in lava flows. Such volcanoes are built of alternating layers of lava and ash.

When lava is very pasty it does not flow away from the vent at all, but heaps up forming a dome covering the vent. The lava may be so viscous that a pillar or spine of hot lava is pushed through the vent. The spine of Mont Pelée which formed in 1902–1903 grew to a height of 300 metres.

Rock formations derived from the products of ancient volcanic eruptions and the eroded remains of extinct volcanoes form the landscape in many countries. Fig 66 is an extinct volcano in the eastern Sahara in which the ash-cone has been eroded away, exposing the central plug as a spectacular spire. Many mountainous parts of Britain are built of old volcanic rocks, including lava flows, *tuffs* (hardened ash deposits), *ignimbrites* (products of glowing clouds, p 36) and pillow lavas erupted under water.

A.PESCE

66 Eroded extinct volcano of Ehi Goudroussou, Tibesti, Sahara

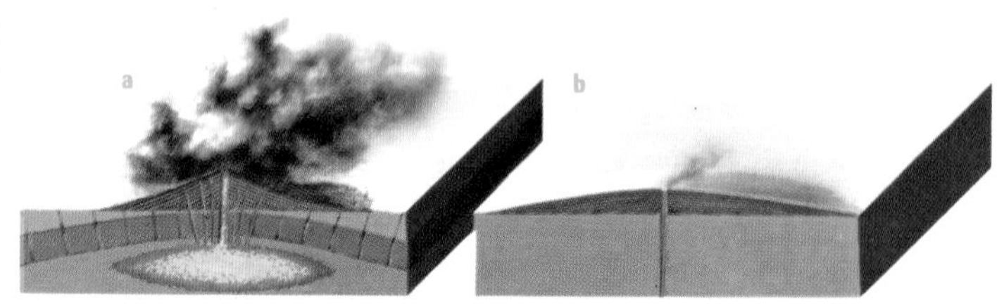

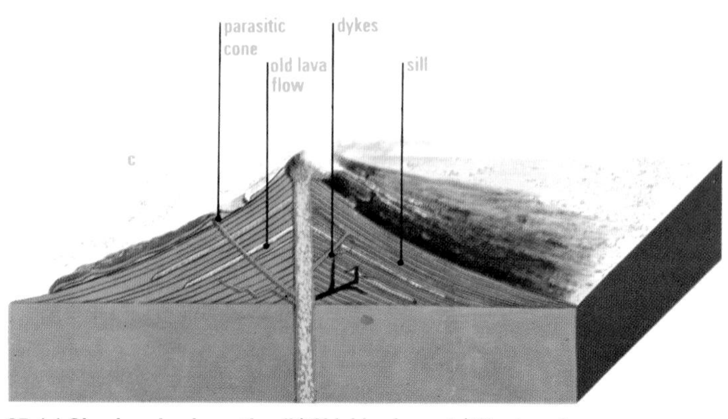

67 (a) Glowing cloud eruption (b) Shield volcano (c) Strato-volcano

ISLAND PICTURES

68 A new flow of smooth-surfaced lava *(pahoehoe)* covering old, rough-surfaced lava *(aa)*, Hawaii

IGS

69 Nuée ardente eruption, Mt Pelée, Martinique. 1902

A.LACROIX

70 Spine of Mt Pelée from ruins of St Pierre

JAMES STEWART

71 Villarrica, Chile: a perfect strato-volcano

Volcanic eruptions

Volcanoes are normally active for short periods, a few months or a year or two at a time. Between eruptions they remain dormant for longer periods amounting to several years or even several hundred years. During this time the lava in the crater solidifies, plugging the volcanic vent. Gas pressure continuously builds up during the dormant period till there is an explosion which opens up a new vent; this is the beginning of a new eruption. A volcano is said to be extinct when it has not erupted in historic times. In the past, volcanoes which were thought to be extinct have erupted again, like Vesuvius in AD 79. From the earliest recorded times, Vesuvius had remained dormant: the fertile soil of the cone was cultivated to the summit and many people lived on the mountain slopes. Earthquakes which took place between AD 63 and AD 79 were not recognised as a warning of the impending eruption, because people were certain Vesuvius was extinct. An enormous explosive eruption shattered much of the cone into fragments and buried the area round about with ash. A thick fall buried the town of Pompeii, killing some people, although many had escaped. Nearby Herculaneum was destroyed by an avalanche of volcanic ash turned to mud by torrential rain from the volcanic cloud.

The most violent volcanic eruptions are associated with pasty lava very rich in gas. Explosive expansion of this gas pulverizes the lava, producing billowing ash-filled clouds. In some eruptions, fragments suspended in a red-hot cloud of lava gas flow downhill, directed by the blast of the explosion, at speeds up to 250 km per hour. These clouds are called *nuées ardentes,* or glowing clouds. The first recorded *nuée ardente* was from Mont Pelée on the island of Martinique in the West Indies, in 1902. The town of St Pierre with 30 000 inhabitants was destroyed by the cloud, which had travelled 10 km from the crater in less than five minutes. Fig 69 shows scientists examining the remains of St Pierre a few weeks after the disaster, looking with apprehension at the approach of yet another *nuée ardente*. The gas-rich *nuée ardente* eruptions continued for some months; later when gas pressure was almost exhausted, pasty lava was pushed up, forming a spine (fig 70).

Mont Pelée shows that a volcano can behave in different ways during different phases of an eruption. Also, as it gets older, a volcano may progressively change its habits. A volcano which starts life as a basalt shield volcano, erupting frequently, may change so that it erupts less often but more explosively with pasty lava, building a strato-volcano on top of the older shield volcano. As a volcano grows older, its structure becomes more complex. The cone builds higher and higher, and greater gas pressure is needed to push magma up to the summit crater. Often it happens that lava does not reach the summit crater, but breaks out on the side of the volcano, forming parasitic cones.

72 Dead Horse Point, Canyonlands National Park, Utah, USA.